Transitions to Sustainability

François Mancebo • Ignacy Sachs
Editors

Transitions to Sustainability

Springer

Editors
François Mancebo
IATEUR IRCS
University of Reims Champagne-Ardenne
Reims, Marne, France

Ignacy Sachs
CRBC – Mondes américains
EHESS, Paris, France

ISBN 978-94-017-9531-9 ISBN 978-94-017-9532-6 (eBook)
DOI 10.1007/978-94-017-9532-6
Springer Dordrecht Heidelberg New York London

Library of Congress Control Number: 2014956565

Printed on acid-free paper

Springer is part of Springer Science+Business Media (www.springer.com)

Contents

Contributors

Jon Marco Church IATEUR-IRCS, Rheims University, Reims Cedex, France

Christian Comeliau Geneva, Switzerland

Arthur Lyon Dahl International Environment Forum, Geneva, Switzerland

Marc Dijk ICIS (International Centre for Integrated assessment and Sustainable development), Maastricht University, Maastricht, The Netherlands

Ladislau Dowbor Faculdade de Ciências Econômicas, Pontifícia Universidade de São Paulo, São Paulo, SP, Brazil

Peter M. Haas Professor of Political Science, University of Massachusetts Amherst, Amherst, MA, USA

Sylvia Karlsson-Vinkhuyzen Public Administration and Policy Group Wageningen University, Wageningen, The Netherlands

Carlos Lopes Economic Commission for Africa, UNECA, Addis Ababa, Ethiopia

François Mancebo IATEUR IRCS, University of Reims Champagne-Ardenne, Reims Cedex, France

Bernard Pecqueur UMR PACTE CNRS, Université Joseph Fourier Grenoble, Grenoble, France

Ignacy Sachs CRBC – Mondes américains, EHESS, Paris cedex 13, France

Paolo Freire Vieira Universidade Federal de Santa Catarina, Campus Universitário Reitor João David Ferreira Lima Trindade, Florianópolis, Santa Catarina, Brazil

List of Figures

List of Tables

Introduction

François Mancebo

This book considers the conditions of transition to sustainability: how to take into consideration new global phenomena such as and of the dimension of climate change, the depletion of natural resources, financial crises, demographic dynamics, migrations and mobility, while bearing in mind short-term or local place-based issues, such as social justice or quality of life. The Millennium Declaration proclaimed the "collective responsibility to uphold the principles of human dignity, equality and equity at the global level." Of course, but how to go beyond lip service and do it concretely?

In 2000, Paul Crutzen and Eugene Stoermer stated that we have entered, since the beginning of nineteenth century and the Industrial Revolution in Europe, a new period of the Earth's history, which they called anthropocene.[1] A period when human activity becomes the main factor that determines the state of the planet, from its biosphere to its land, from its climate to its seas. Indeed, since 2009, a working group of the International Commission on Stratigraphy is considering making the anthropocene officially a geological epoch. How to better highlight the responsibility of human societies toward the "spaceship Earth," to use an expression first employed by Kenneth Boulding?[2]

François Mancebo is a Professor of Planning and Sustainability at Rheims University. He is the Head of the IRCS (International Research Center on Sustainability) and the Director of the IATEUR (Institute of Regional Development, Environment and Urban Planning of Rheims). Email: francois.mancebo@univ-reims.fr.

[1] Crutzen P. J., Soermer E. F., 2000, "The Anthropocene," *Global Change Newsletter*, n° 41, pp. 17–18, IGBP.

[2] Boulding K., 1966, "The Economics of the Coming Spaceship Earth", *Environmental Quality in a Growing Economy,* Boulding K. et al. eds., pp. 3–14, John Hopkins University Press.

F. Mancebo (✉)
IATEUR IRCS, University of Reims Champagne-Ardenne,
Campus Croix-Rouge, 57 bis avenue P. Taittinger, 51100 Reims Cedex, France
e-mail: francois.mancebo@univ-reims.fr

F. Mancebo, I. Sachs (eds.), *Transitions to Sustainability*,
DOI 10.1007/978-94-017-9532-6_1

Since the United Nations Conference on Environment and Development (UNCED) of 1992 in Rio, when "sustainable development"—defined in *Our Common Future*[3]—was given an operational framework, the notion of sustainable development successfully spread in the political world and more largely within the civil society. So much so that, when another Earth Summit was organized again in Rio 20 years after—in 2012—its new name was "United Nations Conference on Sustainable Development".

These last 20 years, the conscience of the environmental, economic and social challenges of our planet and its inhabitants has greatly evolved, while the geographical and political context changed dramatically. Besides, new actors emerged on the sustainability policy scene—firms and companies, NGOs, local communities, etc.—alongside the traditional institutional actors such as states and international organizations. The current institutional framework for sustainability is not really able to take charge of this new configuration. It is not an accident that one of the two major topics of Rio+20 was "the institutional framework for sustainable development." What could be a new policy framework to foster efficient transitions to sustainability? Indeed, the recurring issue of coordination mechanisms—be it at local, national, regional or global level—is a crucial one. One thing is already obvious: transition to sustainability demands serious changes in the way humans do business with each other and with the earth, in the face of a fractured, unequal world.

Thus, one question that need a clear and complete answer is: How to link social justice with sustainability policies? What governance tools to do so? Engaging which parties? Many environmental problems—climate change, land degradation, urban sprawl or loss of biological diversity, to list just a few very different issues—receive first inefficient answers under the double pressure of the divergent political agendas of the different actors and of the lobbying, and then knee-jerk panic reaction treating symptoms not disease, when the public opinion start demanding immediate action. But this way of doing—especially its panic component—is very poor policy. For example, the systematic recourse to environmental technologies to meet with sustainability issues is typical of such final knee-jerk reactions. Like a deceiving Promethean promise, it trades the hope to combining successfully environmental improvement with economic growth, against huge unforeseen real environmental and social side effects. Solutions to problems can create problems of their own.

If there is often critical need for rapid transition to sustainability in different matters, urgency itself brings a risk of short-termism and inappropriate reaction. In 2000, the Millennium Declaration proclaimed: "We recognize that, in addition to our separate responsibilities to our individual societies, we have a collective responsibility to uphold the principles of human dignity, equality and equity at the global level." Today, on the verge of a new cycle, where Sustainable Development Goals will replace the former Millennium Development Goals, it is crucial to go beyond the mantra and ask: how to link practically long-term and short-term priorities, place-based and global approaches, traditional institutional actors and local communities interests? Meeting this challenge requires an inclusive approach of sustainability,

[3]WCED, 1987, *Our Common Future*. Oxford University Press.

where the societal processes of change related to the emergence of new actors, the adoption of new collective behaviors and the definition of new social representations—socially shared meanings—are of considerable importance.

The challenge, here is introducing social innovation as a key factor for a sound transition to sustainability: In other words, it is a matter of designing a new social contract that includes—in the footsteps of regretted professor Elinor Ostrom—communities of interest, neighborhood communities and groups of individuals forming voluntary associations, among the main stakeholders of sustainable development. Determining the conditions and the form of this new social contract is crucial, since it has a lot to do with defining what a "good" environment means. Transition to sustainability requires more than developing the right markets, institutions and metrics. It requires social momentum—a social movement for change. In this sense, transition to sustainability can be conceived as a long-term democratic project, taking place simultaneously in its social, environmental and spatial dimension. Which kind of society do we want to live in? Who decides on necessary compromises? What control and validation methods are possible? Which compromises between the goals and interests of the different groups? What linkage between one decision-making level and the other? These questions are major issues to design sound transitions to sustainability. To try bringing an answer, the book is organized in three parts.

The first one—Meeting the Challenges of the Anthropocene: Back to planning?—aims at identifying new form of planning, which could foster transition to sustainability. Christian Comeliau considers that this planning should be designed as a political process rather than just a technical or economic program, since what is at stake here is nothing less than the type of society we are going to promote in the long run. Ladislau Dowbor, who addresses a new form of planning in Brazil—which he calls economic democracy—develops this approach in a place-based context. Meanwhile, at the international level, Ignacy Sachs proposes the elaboration of a 15-year world development plan for the period of 2016–2030. But Jon Marco Church reminds us that the international system is anarchical. It may be somewhat premature to imagine right now an efficient world development plan. According to him, at this stage the question still is: Can sustainability planning be considered as an emerging norm at the international level?

Indeed, norms embody values and ideals. Thus their emergence may also promote a new social contract. Thus, defining a new form of planning fits into a larger picture, which is the subject of the second part of this book named "Towards a New Social Contract". According to Carlos Lopes, a collective law embodied in a social contract makes a lot of sense when addressing transition to sustainability. According to him, the main aspect of this contract is that present generations are held accountable by future generations. How to do this? How to design a more sustainable future? Peter Haas considers two strategies: Harnessing shared norms and causal beliefs behind a direct sustainability agenda, or aggregating different agreements out of the hope that the whole will be larger than the sum of its parts and will give rise to a second order sustainable transformations. But whatever the strategy, how to determine if an adopted pathway to sustainability is successful or not? What indicators should be considered? Arthur Dahl proposes indicators of well being, including material, social, cultural and spiritual dimensions of human progress that would highlight disadvantaged minorities, gender and class differences, and other priority needs of specific populations. In the same spirit, François Mancebo addresses in his chapter—Insights for a Better Future in an Unfair World—one the more challenging aspects of transition to sustainability: combining sustainability policies with social justice. Since sustainability programs may turn out to be completely out of touch with the needs and expectations of the populations concerned, he proposes to address the social process of decision-making itself by promoting people's place-based appropriation of sustainability policies.

The third part—Some Governance Issues—addresses global energy governance, multi-stakeholder governance for sustainable mobility, and territorial governance. Sylvia Karlsson-Vinkhuyzen analyzes the present lack of legitimacy of global energy governance. She demonstrates that strengthening global energy governance is not normative matter but rather a matter of subjective legitimacy in the eyes of governments and other actors. Marc Dijk champions a not so different position on the subject of sustainable mobility. The nature of mobility issues—multi-faceted, involving social, economic, and ecological as well as technical aspects—requires multi-stakeholder governance. In both cases two questions may emerge: Who is invited to the "governance party," and on what geographical and temporal scales? This is what Bernard Pecqueur and Paolo Vieira strive to address with the notion of *territorial governance*, which supposes policies built by multi-actor panels exteriors to the classical politico-administrative structures. The point is empowering local communities, so as to create a real change in their perceptions, attitudes and behaviors. The third part ends with a declaration—Rheims Sustainability Vision—made at the 3rd *Rencontres Internationales de Reims on Sustainability Studies*, as a contribution to the open working group on Sustainable Development Goals.

This book is based on the debates and the outputs of the last three *Rencontres Internationales de Reims on Sustainability Studies*, and international conference organized annually by the IRCS (International Research Center on Sustainability—www.sustainability-studies.org) at Rheims University. The IRCS is engaged—together with other research centers around the world—in sustainability science: An emerging field of research, which objectives are to generate useful knowledge to support a transition to sustainability.[4] Sustainability science considers the interplay and dynamic evolution of social, economic and natural systems, on an integrated and long-term perspective at different geographical scales, from global to local.[5]

---

[4]Clark W. C., 2007, "Sustainability Science: A Room of its Own", *Proceedings of the National Academy of Sciences*, vol. 104, n° 6, pp. 1737–1738

&

Kates R. W., Clark W. C., Corell R., Hall J. M., Jaeger C. C., Lowe I., McCarthy J. J., Schellnhuber H. J., Bolin B., Dickson N. M., et al., 2001, "Sustainability Science", *Science*, vol. 292, n° 5571, pp. 641–642

[5]Swart R. J., Raskin P., Robinson J., 2004, The Problem of the Future: Sustainability Science and Scenario Analysis", *Global Environmental Change*, n° 14, pp. 137–146

Part I
Meeting the Challenges of the Anthropocene: Back to Planning?

Entering the Anthropocene: The Twofold Challenge of Climate Change and Poverty Eradication

Ignacy Sachs

Abstract We are all living in a new era since the beginning of the industrial revolution: the Anthropocene, which reflects the growing, influence of the human activities on the earth. If we are responsible, we must limit our environmental impacts without forgetting that we have a huge social agenda ahead. So, we have to work on a tripod: the goals are social, there are environmental conditions to be respected, and if we do not give economical viability to our project, it will not happened. Economic viability does not come out of nowhere. We are in a period in which planning is totally demoralized and markets do not know how to manage themselves, they are shortsighted. And, in sensitive to social dimension, if we want to reintroduce social dimension, if we want to organize ourselves in a long time perspective, we have to go back to planning.

Keywords Anthropocene • Social inequalities • Social contract • International cooperation • Planning

In Molière's play, *Le bourgeois gentilhomme*, Monsieur Jourdain speaks prose without being aware of it. Similarly, we have already been living for a while in a new geological era – the anthropocene, *the age of mankind* – defined by Paul Crutzen as a result of "the central role of mankind in geology and ecology" (Crutzen and Stoermer 2000).

In reality, the entry into the anthropocene, prompted by the industrial revolution of the eighteenth and nineteenth centuries, marked the second major turn in the extensive history of the existence of our species on the spaceship Earth. The first, the Neolithic revolution (Childe 1942), had started 12 years ago in Mesopotamia. It paved the way for the transition from hunting and gathering to agriculture and

Ignacy Sachs is Honorary Professor of Development Economics at the EHESS (French school of advanced social studies). He acted as a Special Advisor to the Secretary-General of the UN Conference on Human Environment in Stockholm in 1972 and of the UN Conference on Environment and Development in Rio in 1992. He developed the notion of eco-development. Email. isachs@msh-paris.fr.

I. Sachs (✉)
CRBC – Mondes américains, EHESS, 190-198 avenue de France,
75244 Paris cedex 13, France
e-mail: isachs@msh-paris.fr

F. Mancebo, I. Sachs (eds.), *Transitions to Sustainability*,
DOI 10.1007/978-94-017-9532-6_2

permanent human settlements out of which the first towns arose. The next breakthrough came with the unification of the world economy as a result of the discovery of America by European navigators in the fifteenth century.

The Twofold Challenge of the Anthropocene

Paul Crutzen rightly insists on the increasingly determinant impact of human activities upon the biosphere: "*Without major catastrophes like an enormous volcanic eruption, an unexpected epidemic, a large-scale nuclear war, an asteroid impact, a new ice age, or continued plundering of Earth's resources by partially still primitive technology (the last four dangers can, however, be prevented in a real functioning noösphere) mankind will remain a major geological force for many millennia, maybe millions of years, to come. To develop a world-wide accepted strategy leading to sustainability of ecosystems against human induced stresses will be one of the great future tasks of mankind, requiring intensive research efforts and wise application of the knowledge thus acquired in the noösphere, better known as knowledge or information society. An exciting, but also difficult and daunting task lies ahead of the global research and engineering community to guide mankind towards global, sustainable, environmental management*" (Crutzen and Stoermer 2000). As a matter of fact, we can no longer postpone the urgent need to tackle two major and closely intertwined challenges:

On the one hand, we need to put an end to the scandal of abyssal inequalities in the living conditions and quality of life prevailing today between nations and within nations, so as to eliminate the scandal of hunger and misdevelopment: a small minority occupying the spacious and comfortable cabins on the deck of the spaceship Earth, while many more are condemned to lead a busy, yet miserable, life in its hold, working hard to survive on a hand-to-mouth basis. Do you recall Fellini's masterpiece *E la nave va*?

On the other, we ought to simultaneously prevent climate change – the warming produced by massive emissions of greenhouse gases that will have deleterious consequences on the living conditions on the spaceship Earth.

By the middle of the twenty-first century, the spaceship Earth will have a population/crew of nine billion, compared to only one billion in 1800; two in 1927, four in 1974; and six in 1999 (see Fig. 1). However, this rapid expansion of the world population should not be seen as a reason for despair, but we ought to slow down the demographic expansion by providing better life conditions to all the passengers of the spaceship Earth, and thus to stabilize the world population around ten billion before the end of the century.

The scientists of the International Panel on Climate Change are adamant: the average temperature should not increase by more than 2 °C if we want to avoid a disastrous worsening of the living conditions for our species.

Is it too much to expect that the anthropocene becomes, on the contrary, the "age of empathy," thus improving on a lasting basis the human condition? For this, it is urgent to learn how to walk on two legs, combining social justice with environmental prudence.

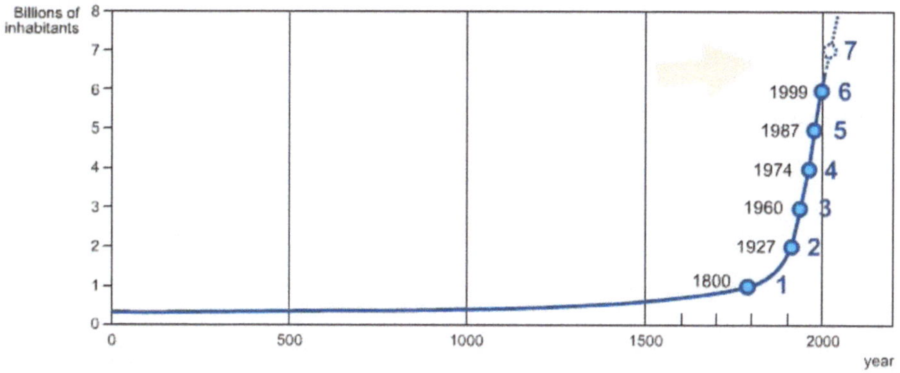

Fig. 1 World population increase in the last 200 years (Source: Institut national d'études démographiques)

To do so, we can no longer rely on the myth of the invisible hand and self-regulating markets. Left to themselves, markets have proved shortsighted and, what is more, insensitive to unbearable social costs. That is why we must urgently assume an active role as *geonauts-planners* (Orsenna 2005) striving to steer, as far as possible, the unfolding history, and a very tall order indeed, if we are to avoid excessive voluntarism. Geonauts-planners will have to find the way between Scylla and Charybdis, in our case, between the predicament of social inequality and the menace of climate change threatening to play havoc with the very survival of mankind.

As a matter of fact, if we want to promote long term strategies governed by care for all humans – present and future – we must turn to the Visible Hand and its five fingers: long term democratic planning, a renewed *contrat social*, food and energy security – the two pillars of inclusionary[1] and sustainable development – and international cooperation.

Back to Planning

The way out may be narrow, but it exists. The biblical needle eye is here an appropriate image. We cannot afford any longer, as already said, to be subjected to the excessively high social costs of adaptation through the free interplay of market forces, hence the *urgent need to institute planning*, both at national and planetary levels. Planning, a child of war economy, was born at the age of the abacus, and in spite of this technical limitation, it proved quite useful in the past. We thus have a

[1] I prefer the term used by A. K. Sen, "inclusionary" to "inclusive," for an obvious reason: inclusionary denotes a movement.

reasonable chance of achieving a much better performance with computers in our hands, even though they are in themselves a useful, but by no means a sufficient condition for efficient planning.

Paradoxically, planning is unpopular in the computer age. There are two reasons to it: one is the die-hard myth of the invisible hand, still convenient to the capitalists, reinforced by the fall of the Berlin Wall in 1989 and the ensuing implosion of the Soviet Union.

The second refers to the sometimes-spurious relation in the past between planning and autocratic regimes. To take an extreme case, in his essay *L'île aux cannibales*, Nicolas Werth describes the attempt by the Soviet planners to populate an island in the middle of a Siberian river with prisoners randomly grabbed in the streets of Moscow and left to themselves in a hostile environment without adequate equipment or technical advice (Werth 2006). Eager to incorporate new territories to the Soviet economy, the planners ended up provoking a return to cannibalism for real, reminding us of a satirical piece by Swift (1729).

It is thus essential to keep one condition in mind: planning only makes sense so long as it is performed within a truly democratic regime, which respects the right to bona fide error. The failures of authoritarian planning in the twentieth century were to a large extent due to the absence of this condition. Too many politicians in power pretended to have the monopoly of truth, and, therefore, the right to arbitrary and often brutal dismissal of dissenting opinions. Truly democratic planning cannot prosper without free exchange of ideas. The least one can say is that this condition was not always respected in the Soviet Union and people's democracies.

Historians specialized in *histoire immédiate* are still to give us a thorough critical evaluation of the "short twentieth century" and its two terrible World Wars (Hobsbawn 1994): the rise and fall of real socialism in the Soviet Union, the emergence of the Welfare States in the Scandinavian countries and of the New Deal in the United States, as well as the diverse models of mixed economies in the developing countries.

Such a study would certainly contribute towards elaborating new paradigms of democratic planning and economic governance and towards designing development strategies, capable of giving the present and future generations a fair chance to make the most of their lives by granting them decent material conditions, hand in hand with the exercise of basic freedoms.

In particular, the patterns of the public sector in mixed private-public developing economies may be analyzed using two historical models that define the range of possibilities: the Japanese model of the Meiji era, in which the State limited itself to play a transitory initial role, and the Nehruvian Indian model, also known as the "socialistic pattern of society" (Sachs 1964).

We ought to open a discussion on possible "third ways" differing from both the classical capitalism and real socialism as we have known them, focusing on mixed public/private economies with a significant third sector of social economy. Neither history (nor planning) should be forced into the straight jacket of inherited categories. There are at least three reasons to believe that the future will differ from the past: the lessons that we shall learn from the historical experiences analyzed *sine ira et odio*, the scientific and technical progress and human creativity at large.

In this context, we should remember that, as recalled by Anatol Rapoport, we are the only living species capable of imagining and anticipating alternative futures (Rapoport 1974), and therefore, of planning, so long as we accept M. Kalecki's succinct definition of the latter as "*variant thinking*" about future action.

Not that we are, or ever will be, "masters of nature" as Descartes wanted us to believe. The nuclear disasters of Chernobyl and Fukushima on the one hand, the tsunamis, typhoons and other natural catastrophes, on the other, should teach us modesty. My preference goes to another seventeenth century French philosopher, Blaise Pascal, who compared man to a "thinking reed" (*roseau pensant*), capable of a twofold strategy: bending under the wind in a *reactive posture*, while adopting at the same time a proactive attitude, in the attempt to outwit nature.

Democratic planning, predicated on a quadripartite dialogue between planners, entrepreneurs, workers and the organized civil society, and making good use of the computers is yet to be firmly established on the tripod of social and ethical goals, environmental conditionality and economic viability, the latter being the *sine qua non* condition for things that matter to happen.

Planning implies an iterative process involving actors at the local, regional, national and international levels. It should incorporate, on the one hand, the concepts of *ecological footprint and* biocapacity, making a sharp distinction between countries which are biocapacity debtors and those which are biocapacity creditors[2] and, on the other, the definition of opportunities for *decent work*, as proposed by the International Labor Organization, and which includes *employment and self-employment*, the latter particularly important in rural societies.

At the local level, following the example of the French Revolution, we need to start by compiling *cahiers de doléances*: a comprehensive listing of the social priorities that must be addressed, side by side with the identification of the potential local resources, the bottlenecks to be removed and the necessary critical inputs to be brought from outside – knowledge, equipment and finances.

Planners diverge with respect to the time horizon of long term planning. The longer the time span covered, the greater the margins of freedom, but also of uncertainty. Twenty years seem to be a fair choice with the possibility to extend further some projections, in order to identify the emerging breaking points.

Essentially, planning is an exercise in systems approach aimed at identifying ambitious, yet feasible, social goals by proposing efficient patterns of resource use while matching them with the available working force by creating opportunities for decent work, so as to eliminate unemployment and, as far as possible, underemployment. For that, planners should address such issues as easing up bottlenecks while avoiding the unnecessary piling up of stocks and turning waste into wealth (a catchphrase in Maoist China) by finding productive uses for residues.

---

[2]According to the Global Footprint Network *2010 Annual Report*, http://www.footprintnetwork.org, in 2011, the world population as a whole was a debtor, having exceeded the available biocapacity by 35 %.

We are still far out from having satisfied the basic material needs of the entire human crew of the spaceship Earth. Attacking social inequalities, both at national and international levels, is thus more urgent than ever, keeping in mind the obvious truth whereby in a finite planet we cannot envisage an unlimited growth of material output. The sooner we reduce the still prevailing social disparities in material consumption standards across the world, the sooner we shall be able to move from a growth economy to a steady state economy. We are certainly not there yet.

What Social Contract?

Le hasard parfois fait bien les choses, let us recall that 2012 marked the third centenary of the birth of Jean-Jacques Rousseau and the 250th anniversary of the publication of his seminal *Contrat social*. It is up to us to show that we no longer accept to act along the principle of *homo homini lupus*, nor do we tolerate any longer the deepening of social inequalities between and within nations. The future belongs to *explicit social contracts* established, both at the national and international levels.

The fundamental question to be asked in this respect is: *how much is enough*? Gandhi replied by saying that "*Earth provides enough to satisfy every man's need, but not every man's greed*." However, we should not carry too far his postulate of voluntary simplicity, even though material over-consumption is by no means an indicator of happiness. We should strive to provide each of the nine to ten billion human beings that will be sailing on the spaceship Earth in the second half of this century with a reasonable income, guaranteeing decent material standards of living, on the understanding that the paramount goal is "*a civilization of being in the equitable sharing of having*," in Louis Joseph Lebret's terms. In other words, we should learn to self-control our material consumption.

As there is no reason to believe that we have exhausted the potential of technical progress and of better use of available energies, an ever smaller parcel of the working time of human societies will be required in the future to produce the necessary material goods. It will therefore be possible to gradually reduce the relative share of societal time ascribed to the activities of the *homo faber*, making more time available for the *homo ludens* (Huizinga 1955). At the same time, we ought to ensure that the total working time and earnings from work are fairly distributed among all potential workers, so that the scourge of unemployment is eliminated (Aznar 1993).

The sky is the limit to cultural and ludic activities, so long as we learn how to make appropriate use of the time freed from work. Keynes was right to warn us, as early as in 1930, against a "general nervous breakdown" that might be caused by such a fundamental change in the social fabric (Keynes 1963 (1930)). We might take example from a tribe living on Asian seashore. Its best artists were invited to compete by making beautiful drawings on the humid sand of the beach, bound to disappear with the next high tide. I am *still looking for a better example of material désintéressement.*

In order to move towards a less polarized world society, we must bring back to the fore the social and institutional reforms which no longer attract the attention they deserve in the national and international agendas. The unfinished land reforms certainly belong to this category, side by side with urgently needed fiscal reforms. The experience of the New Deal in the United States ought to be revisited in the latter context.

Food Security

Food and energy security ought to be considered as the two pillars of socially inclusionary and environmentally sound development strategies.

Even assuming a stabilization of the world's population by the mid twenty-first century, we must think of how to provide enough food and a reasonably comfortable life for 9+ billion human beings. For that, we depend on further progress of the green and blue revolutions with special reference to the interface between the two and to their dissemination throughout all the continents.

Rather than sticking to the elitist green revolution as initially proposed by N. Borlaugh, based on the massive use of selected seeds, fertilizers and water, we must move towards the "*evergreen revolution*," as advocated by the well-known Indian agronomist M.S. Swaminathan[3] and directed towards the small peasants who still represent the majority among rural dwellers in developing countries (Griffon 2006).[4]

At the same time, we ought to limit the devastating impact of extensive cattle breeding on forests by resorting to a husbandry better integrated with small scale family agriculture as well as encouraging the substitution of meat consumption by that of fish coming from pisciculture.[5]

This leads us to emphasize the importance of intensive production units combining horticulture, pisciculture and arboriculture inspired by the traditional dike pond systems in Southern China (Ruddle and Zhong 1988). Such systems can also be adapted to different Brazilian contexts – the Amazonian *igarapés*, natural and manmade lakes, ponds and *açudes*, as well as the coastal areas, in particular the lagoons behind the reefs.

---

[3] The M. S. Swaminathan Research Foundation is a reference with respect to sustainable agriculture and rural development (http://www.mssrf.org/bd.html).

[4] See on this point, Bruno Parmentier who opposes the two green revolutions in the following terms: "*Là où la révolution verte cherchait à artificialiser le plus complètement possible le milieu naturel, la révolution doublement verte vise à inscrire le système productif au sein des écosystèmes. La première force la nature via le recours massifs à des intrants, la seconde l'accompagne en recherchant un équilibre entre potentiel interne et apports extérieurs; la première spécialise les productions, la seconde les diversifie pour qu'elles se renforcent mutuellemen t; la première recherche une protection absolue de la production via l'éradication complète des maladies et des ravageurs, la seconde compose et gère le système pour contenir ces envahisseurs*" (Parmentier 2007, p. 160).

[5] As far as Brazil is concerned, fish farming has a great future in the Amazon region, the Mato Grosso Pantanal, not speaking of the Atlantic coast, so long as a satisfactory solution is found for long distance transportation of fish, more efficient than by road, and cheaper than by plane. Should we give a second chance to the zeppelin, the more so that we can fill it today with non-inflammable gas?

The following diagram illustrates such a unit (Fig. 2):

Two hundred people fed throughout the year on half a hectare, too good to be true! If it were only possible to create throughout Brazil an archipelago of one million of such units, it would ensure the food security to 200 million Brazilians and furthermore, generate 2.5–5 million jobs, while preventing further deforestation. Only floriculture can compete with such units as far as employment per hectare is concerned, but obviously, the demand for flowers is much more limited than that for food.

At any rate, we are not there yet, and we ought to realistically evaluate the prospect of advancing along these lines. But the challenge is certainly worth a try. Brazil and India could well join their forces in this endeavor.

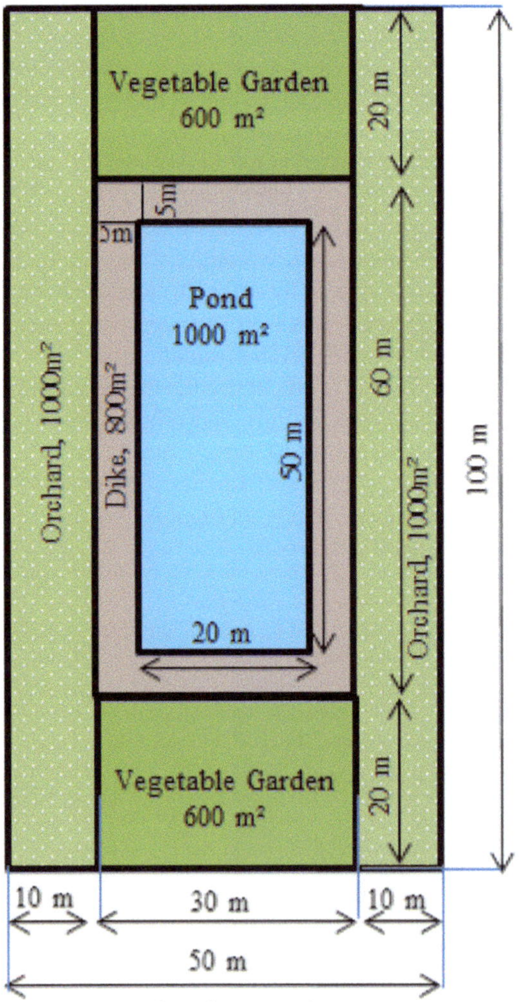

Fig. 2 Dike-pond-system in Southern China
A 1,000 m² pond (50 m × 20 m) producing 10 tons of fish (a productivity deemed feasible by the BNDES), corresponding to the yearly consumption of 200 inhabitants
A 800-m² dike and an additional surface of 1,200 m² used for vegetable gardens meeting the annual requirements of 200 inhabitations (feasible on the condition of resorting to biochar as advocated by the NGO Pro-natura international at the rate of 1 kg/m²)
A 2,000 m² surface for arboriculture

Energy Security

Let us start by a truism: energy is crucial to development insofar as it increases the productivity of human work. The industrial revolution was predicated on resorting to a widespread use of coal and later oil and gas, three fossil energies responsible for the emissions of carbon dioxide and the ensuing global warming. That is why we ought to reduce and even phase out the use of fossil energies even before they are entirely exhausted.[6]

Fortunately, we may shift to an array of alternative energies: solar, hydro, wind, geothermal and biomass, each one presenting some advantages and obstacles to be overcome. Three remarks are in order here:

- The energy strategy should address three interrelated questions: energy sobriety, efficiency, and alternative energy sources (Dessus 2011). Profligate use of energy can be curved and efficiency greatly increased, so as to reduce the final demand for energy.
- Resorting to bioenergy calls for a careful evaluation of the potential conflict for scarce land resources between the production of food and that of energy. It need not happen if residues from food production are used as a feedstock for energy production (cellulosic ethanol, biogas from cattle dung, etc.). Moreover, countries like Brazil have enough agricultural land available to still afford expanding both food and bioenergy production, so long as this does not happen at the expense of standing forests. The latter are to be conserved as carbon sinks, not speaking of their other potential economic uses. Algae grown in sea and fresh water are also a potential source of bioenergy. Food, animal feed, fertilizers, fibers (standing for all kinds of industrial feedstock), fuel and standing forests are the six potential uses of biomass, which ought to be articulated through the development plan.
- The Faustian bargain – resorting to nuclear energy – poses a serious dilemma. It is clean in terms of emissions of carbon dioxide, however, not immune to the risk of highly improbable, yet devastating accidents, not to mention the danger of the proliferation of nuclear weapons. That is why some countries have recently taken the wise decision to phase out the production of nuclear energy (Dessus and Laponche 2011).[7]

---

[6] This is not tantamount to abstaining from the exploration of the offshore "pré-sal" oil reserves in Brazil. However, utmost attention should be given to the prevention of potentially dangerous ecological accidents. On the other hand, the "pré-sal' oil could be taxed so as to finance the phasing out in future of fossil energies.

[7] The French president, François Hollande, proposed in his electoral programme to reduce the relative share of nuclear electricity from 75 to 50 % by 2025 (*Le Monde*, 17/12/2011). The French Green Party favors a total phasing out of nuclear energy.

Conclusions: About International Cooperation

The UN system has a major role to play. On the one hand, it should greatly increase its financial participation in assisting the less developed countries in their socially inclusionary and environmentally sustainable development. For that, the UN could rely on the following funding:

- A contribution to a UN Development Fund from the developed countries, going back to the much discussed but never implemented pledge of allocating annually for this purpose at least 0.7 % of their GNP (Laget 2009).[8]
- The proceedings from the Tobin tax to be collected on financial transactions.
- A tax on carbon leading to the reduction of carbon emissions and used to finance projects that meet this goal.
- Tolls on oceans and air as a fee for using parts of mankind's common heritage, with the possibility of exempting on certain conditions ships and planes from less developed countries.

It is difficult to set long term quantitative goals but it should not be impossible to aim globally at 2 % of the world GNP, a very large sum indeed, if it were to be used to increase the productive investment and social expenditure in developing countries.[9]

On the other hand, the UN system should use its expertise and organizational skills to create international networks for scientific and technical cooperation among countries sharing similar biomes, instead of using the geography of proximity. This will call for a significant overhaul in the organization of the UN affiliated bodies, requiring in particular, a much closer cooperation between regional commissions (ESCAP, ECLAC, ECA, ECE) substantive agencies (FAO, UNIDO, UNESCO, UNDP) and financial institutions (the World Bank, the regional and national development banks), around biome-oriented programs for humid tropics, semi-arid regions, and savannas, temperate regions and so on. Without forgetting the crucial interface between water and land ecosystems in which the green and blue revolutions interact in the intensive production units described above, combining horticulture, pisciculture and arboriculture.

In particular, the UN member countries should be invited to present within 2 or 3 years national long term development plans containing the relevant information about the ecological footprint and biocapacity use, as well as about social objectives and especially the creation of opportunities for

[8] In 2007, the OECD contribution was of 0.28 %. Only five countries contributed in excess of the 0.7 % target: Norway (0.95 %), Sweden, Luxemburg, the Netherlands and Denmark. The French was of 0.38 %.

[9] In 2011, the global GNP reached 70 billion US dollars at market exchange rates and 79 billion US dollars at purchasing power parities (IMF 2012)

(continued)

decent work. The biocapacity debtors should be encouraged to decrease their ecological footprint while the biocapacity creditors should be assisted in making better use of their biocapacity. Another urgent question is the choice of the energy paradigms.

In parallel, the UN should advance in expanding the international capacity to finance development and setting a collaborative network aimed at designing and implementing biome-based development strategies, which imply much closer South-South links and collaboration along parallels rather than along meridians.

In this way, conditions would be created to move to the next stage of international cooperation: identifying the synergies that can be achieved by mutually adjusting the national development plans, opening new opportunities for trade as well as for scientific and technical exchanges, to be partly financed from international sources and assisted by the network of UN agencies.

In this way, the ground might be prepared to move towards the elaboration of a *first 15-year world development plan* for the period of 2016–2030, to be followed by a second *20 year world plan* for the 2031–2050 time span. Technically, such world plans are within reach. What is still missing is the political will and the initiative to move as quickly as possible in this direction. It has taken us more than two centuries to acknowledge the dawn of a new age. We must now accelerate the pace of the required transformations to make up for the delay, so as to successfully enter into the Anthropocene.

References

Aznar G (1993) Travailler moins pour travailler tous. Syros, Paris

Childe VG (1942) What happened in history. Pelican Books, London

Crutzen P, Stoermer EF (2000) The 'Anthropocene'. Glob Change Newsl – The International Geosphere – Biosphere Programme (IGBP): A Study of Global Change of the International Council for Science. 41:17–18

Dessus B (2011). Scénario négaWatt 2011. Association négaWatt – www.negawatt.org

Dessus B, Laponche B (2011) En Finir avec le Nucléaire – Pourquoi et Comment. Éditions du Seuil, Paris

Griffon M (2006) Nourrir la planète. Odile Jacob, Paris

Hobsbawm E (1994) The age of extremes: the short twentieth century, 1914–1991. Penguin Books, London

Huizinga J (1955 (1938)) Homo Ludens, a study of the play element in culture. Beacon Press, Boston

IMF (2012) World economic outlook: coping with high debt and sluggish growth. International Monetary Fund, Washington, DC

Keynes JM (1963 (1930)) Economic possibilities for our grandchildren. In: Essays in Persuasion. W.W. Norton & Co., New York, pp 358–373

Laget P (2009) Les chiffres d'une planète de fous... ou de l'urgence d'un développement durable. Éditions de l'aube, La Tour-d'Aigue

Orsenna E (2005) Portrait du Gulf Stream. Éloge des courants: promenade. Éditions du Seuil, Paris

Parmentier B (2007) Nourrir l'humanité – Les grands problèmes de l'agriculture mondiale au 21e siècle. La Découverte, Paris
Rapoport A (1974) Conflict in man-made environment. Penguin Books, London
Ruddle K, Zhong G (1988) Integrated agriculture-aquaculture in South China: the dike-pond system of the Zhujiang Delta. Cambridge University Press, Cambridge
Sachs I (1964) Patterns of public sector in underdeveloped economies. Asia Publishing House, New Delhi
Swift J (2013 (1729)) For preventing the children of poor people in Ireland from being a burden to their parents or country, and for making them beneficial to the public. Harper Torch
Werth N (2006) L'Île aux cannibales. 1933, une déportation-abandon en Sibérie. Perrin, Paris

Towards a New Development Planning: The Pre-eminence of Political Choices

Christian Comeliau

Abstract As a technique for organizing economic growth and development, development planning – similar to what was experimented by various countries in the decades of the middle of the twenty first century – this kind of planning is dead. Social progress in all societies, however, is more needed in our time than ever. But the market economy alone is not, by nature, capable to solve some of the major issues raised by this development perspective. A new kind of organization, using market mechanisms but including also a new sort of planning, is to be imagined. The present chapter is only a preliminary contribution to this collective research.

Keywords Development planning • Decision-making • Political strategies • Environmental justice • Public policy

Let's start from a paradox. On the one hand, the economic, social, ecological, cultural and political system of our world is facing extraordinary challenges: in spite of considerable scientific and technological progresses in the last two or three centuries, our system does not seem acceptable anymore, for ethical and political reasons (among which the persistence of deprivation, inequality and violence for a majority of the world population); and it does not seem sustainable either, for economic, social and ecological reason (due to the contrast between the ambition for indefinite growth of production and consumption, the lack of economic and social organization for mastering the structure of this growth, and the accelerated deterioration of our natural resources and environment). On the other hand, until recently, we have believed in the capacity of our economic market-oriented system for self-regulation; in addition, the collapse of the "socialist" world and the reinforcement of the neoliberal ideology in the last 30 years have systematically and constantly discouraged

---

Christian Comeliau is Honorary Professor of Development Economics at the *Institut Universitaire d'Etudes du Développement* (Institute of International and Development Studies) in Geneva. He was an economist at the World Bank and also worked at the Commissariat du Plan in Paris, in the 80s.

C. Comeliau (✉)
Geneva, Switzerland
e-mail: christian.comeliau@club-internet.fr

F. Mancebo, I. Sachs (eds.), *Transitions to Sustainability*,
DOI 10.1007/978-94-017-9532-6_3

any public intervention in the realm of the national economies. In this context, the very idea of development planning at the state level (and, a fortiori, at the international level) has almost completely disappeared: the techniques of management of the "command economies" in the socialist countries have been replaced by capitalist, or at least "State capitalist" approaches, even in China; the practices of combining plans and markets in the West have been practically abandoned (in France, for example, where the effort to restore planning approaches at the beginning of the 80 was eventually a total failure); and in the developing world, the main successes (those of the "emergent countries" like Brazil or India) seem to be the result of a remarkable adjustment of those economies to the new opportunities offered by the world economy, rather than the result of the traditional approaches of planning at the national level.

As a whole, the market system seems to be at the same time triumphant, at least for a minority of the world population and in terms of its own criteria (profit maximization and economic growth), and deeply incapable to master its own successes and deficiencies, especially in terms of social equity and sustainability. Or in other words, there are obviously some spectacular winners in the recent economic game at the world level; but the pursuit of the very idea of the "general interest", in what could be a "healthy" society in the world level, given its technological and even organizational capacities, does not seem to lead anywhere along this line: nobody knows what the "general interest" could be at the world level. The present world crisis (maybe especially in Europe) is at the same time the result and the symbol of this failure of our world, even if we still consider that the economy and the economic performances are not the only measure of the health of a society. Beyond all the traditional, technical indicators of "crisis" used by economists, the world society is obviously in full disarray: the present is extremely difficult for many people, the future may be worst, in the short term as well as for the long run, the opinion is becoming more and more anxious, nobody knows where the world is heading anymore. Indeed we seem very far from the hope expressed, by John Maynard Keynes in the 30s, regarding the probability, in a future less distant than a century, of a world delivered from the anxiety of the economic problem (Keynes 1931).

I start form this paradox and from these questions about the nature, and the building, of the "general interest" of the world society, because I believe that it is the main, and the right, dimension for a discussion about a new conception of development planning today, much more important for social progress than any other question of a technical nature. Any effort to restore the old ways of socialist planning, or those of planning in market economies after the Second World War, or of development planning as practiced in the Third World economies in the 60s or the 70s, is probably pointless today, because the world has changed too considerably; but the self-regulating virtues of the market system in the same context are not more convincing any more, if we try to think the world in terms of social and ecological sustainability in the long run. We have to imagine new ways of thinking, and elaborating policies, about the possible content of the general interest in the twenty first century, at the world level as well as at the regional and at the national level, and about the conditions and the means to implement those polices. When I speak about the "general interest", I mean, in fact an image of a more desirable society in the future: again, it is not primarily in terms of economic performance that we have to think about the future, but by imagining the new characteristics of the society that we want to promote in the future. Those characteristics include some general goals (for example, a richer society, or a more open society, or a more equalitarian one); it implies also, inevitably, various kinds of arbitrages or compromises between the specific goals and interests of the various social groups. Let's add that those characteristics should be built in very concrete and practical terms, that is, in relation with some practical needs of the people (housing, security, food, and other basic needs). This "societal approach" of progress is what I would like to call planning, or development planning, in the context of our discussion.

This is, of course, a very ambitious question, and I can suggest only some preliminary arguments for such a global thinking. I shall remind, in the next section, what was called planning recently (let's say, mainly in the 60s, and only with some examples): this brief description shows us very clearly the main reasons why this kind of planning is not sufficient in the present context of the world economy and society, and not compatible with the new requirements for social and ecological sustainability in the long run. In the critique that I would like to suggest, in the following section, my main argument on this point will be that we should focus our planning efforts, much more than in the past, on the choices of development objectives, what I shall call "political choices" as opposed to "economic or technical choices", and on the progressive translation of those objectives in programs of action and practical measures for a given period; I shall briefly examine, from that point of view, the necessary conditions for a more productive dialogue between the decision-makers and the "experts" (economists, engineers, technicians of various sorts). In this line of reasoning, I shall try to identify, in the conclusion, some major characteristics of the new kind of planning that we should develop, while mentioning some of the corresponding foreseeable difficulties.

Development Planning Around the 60s

I propose to base my critique of the traditional approach of what I call "the usual approach of development planning" on some experiences that I know a little better and about which I have tried to make and to publish some comparisons in the past (Comeliau 1999, 2007). This comparison is not quite recent; but as far as I know, there were few original or new experiences in the last decades. More importantly, my comparison is obviously very partial: it is based on a very small number of experiences, and in particular, it does not include at all the "socialist" experiences of the East, especially that of the USSR and of its satellite countries that would require a much different and broader study of what is a "command economy". My own study was based on three kinds of cases: some planning exercises in several African countries since the 60s, where I have been personally involved (for making some preliminary studies, or as an observer for some international organizations); the French experience of planning, at least the part of this experience in which I took part for several years in the 80s (as an economist at the "Commissariat du Plan" in Paris); and finally the one that is probably the richest of all in terms of development, the Indian experience, of which I had the opportunity to be an external observer in some circumstances.

This is not the place to re-expose the main conclusions of this comparison. What I want to do here is only to make a very brief assessment of these experiences and of the contrast they present, from the point of view of what I have just said of development planning, as an approach of the "general interest" and of the characteristics of the society in the future. I would like only to suggest some elements of approval and some elements of criticism, at a fairly general level. Approval because, obviously, in the majority if not in all of these experiences, as well as in the "plans" produced in this context,[1] there is an effort to identify the main issues of the future of the society as a whole, to sketch the main lines of the desirable evolution, and to propose some of the corresponding needed corrections to the actual situation. Criticism because in these experiences, this reflection on the desirable future is in fact only a small part of the whole exercise, and takes only the form, in most cases, of some general and solemn declaration at the beginning of the document of the plan, but without much systematic relation with the rest of the content in the texts of the plans.

What then is the content of the rest of the plans? Again I am obliged here to summaries, and I would like to try not to make a caricature of these long and heavy exercises. But my impression is that, beyond these general declarations, there are basically three kinds of elements in these plans: a macroeconomic framework (which seems sometimes to be – or to be presented as – the more technical, and therefore the most "serious" part of the plan); a sketch of the main orientations of various policies in some specific sectors; and (or sometimes as the main element of the whole exercise) a series of specific "projects", mainly "investment projects" in capital which add up in a 3 years or 5 years budget and are eventually supposed to be the core of the planning exercise. There is, of course, an assumption of coherence between those four components of the plan: the introductory declaration is suppose to provide and explain the general inspiration of the strategy proposed; the macroeconomic approach is a calculation of the economic (or sometimes only the arithmetic) coherence of the exercise in quantitative aggregates terms; the policies and the list of projects are supposed to be the main practical translation of the orientations of the whole plan. The additional hypothesis is made that the priorities given in the declaration of intention and the general goals announced for the development strategies (if any) are compatible with those included or implicit in the specific policies and in the list of projects.

[1] Some observers agreed however, in the past, to keep a distinction between *the process of planning*, which is a general effort to make a diagnosis of an evolution and to prepare policies for improving this evolution) and the final product of *the successive "plans"*, which are periodic translations of this efforts for a given period.

From the point of view of the present study, this conception and this content raise at least the three following questions:

- Where do the general orientations and objectives of the planning exercise and of the planning process come from?
- What is supposed the main function of the macroeconomic framework?
- Is there any precise connection between the general orientations of the whole plan, on the one hand, and the specific policies and the specific projects, on the other, and how is this connection organized and monitored?

In this section, I shall comment mainly these first questions, and we shall come back to the global set of issues in the following section. Regarding the origin of the general orientations of the plan, I have suggested above that, ideally, to be meaningful as an effort for the whole progress and for the general interest of the society, the development process is supposed to emanate basically from a "vision of the future", built for and by the society itself, with its own preferences and with its own arbitrations between national interests, rather than under various pressures or recommendations from any external power (national or international). We can consider also that it seems desirable (at least according to a majority of opinions available on the subject, but not necessarily according to the ambitions of the national governments, international agencies or large private corporations, which most of the time will prefer a "top-down" decision process) that this vision correspond to a common view of a majority of citizens and interest groups within the society, through a "bottom-up process", as democratic as it is possible when one takes into account the traditions and the social structures of the society concerned. Practically, however, the technical complexities of the many decisions to be made, the multilevel decision process involved (at the international, national and local levels), especially as a result of the globalization process of the world economy, and finally the unequal competition between all actors involved in the process (given their unequal power, knowledge and technical capacities), most of the practical planning processes will be the result of the dominance of a "top-down" process led massively by "experts", especially those coming from the economic departments of the national administrations involved, those belonging to international organizations with economic and financial jurisdictions, and those coming from some major private corporations expecting to make some profits, or at least to preserve their privileges, by participating in the public planning exercise.

Given these perspectives, the main practical question in this first category of problems becomes, as a result: what is the nature of the game between experts, what kind of public interest may result from this game (in terms of the heterogeneous interests of the majority of the populations, especially those who are less capable to be listened in this game, as the rural population, for example, or the majority of the workers in the informal sector), and finally are there any way to transform this result in a more democratic orientation? We shall have to come back to those questions, especially in order to identify the nature of the political choices where these considerations matter, and to re-examine the possible role of experts by taking into account this difference between political and technical-economical choices.

The second order of questions mentioned above is obviously in the same line: what is the main function of the macroeconomic exercise, which sometimes seems to be considered, by a majority of partners or observers, as the main component, and also the main determinant, of the planning process? It is clearly needed, but not as a criterion, or as a fundamental reason, or as a determinant of each of the economic decisions to be made: its function is more about identifying the consistency requirements, in the economy as a whole, of the political choices that can be envisaged, and then decided or cancelled. We shall again have to be more precise about this function, in relation with the indispensable but secondary role of experts. And there is not much to be added to the third order of questions, related to connections between the plan as a whole and the specific policies and projects: it is obvious that it should be re-examined in the same perspective.

Preliminary Elements for a Different Process of Development Planning: Three Categories of Questions

This makes a list of themes, or, more exactly of concepts and methodological requirements to be re-examined in this chapter, which, as a result, could be conceived more as a sort of methodological introduction for building a different sort of development planning. This reflection should focus around the following questions, in which we include, of course, the questions identified above, about what we called the usual approach of development planning:

- What is a political choice, what is the main difference between the political choices and the economic or technical choices? Why is this distinction practically important?
- What is the role of experts in the planning process?
- What is the function of the macroeconomic framework?
- What is the nature of the connection between the global process of development planning and the choices of policies and projects?

Let's examine very briefly the essential of these methodological questions.

Political Choices, Economic or Technical Choices

I propose to consider that adopting a "political" vision or a "political" approach about the elements of the organization of a society, especially when looking for the "general interest", or for the "public interest", in this society, means basically three things:

(a) Adopting a global vision of the main components of this general interest, in their interdependence and in their interrelations;
(b) To be concerned primarily with the finalities, the broad goals pursued by this society in its organization effort (what I have already mentioned previously when suggesting that a development strategy should be based on, and organized around, "the main characteristics that appear desirable for this society in the future");
(c) And consequently the need for arbitrations between the interests of the main social groups, within the society, that are involved and concerned by the building of this strategy.

In contrast, and as a consequence of this definition, "economic" and "technical" choices should be defined as the choices of the "ways and means" to be selected and used in order to implement, in the best possible conditions, the finalities and the goals politically chosen by the society.

I perfectly realize that this definition is debatable, and rather crude, and that, in most of the practical cases, the distinction may not be clear, or not even possible, between those two categories. Or more exactly, these two categories of choices are most of the time closely mixed in the series of successive practical decisions that decision-makers are supposed to pronounce. As a result, many of the practical choices (and even probably most of them at a global level) do include simultaneously, in fact, some political elements and some economic and technical elements in their content. In the practice of building a strategy, it is obviously impossible, or at least unrealistic, or even clearly not desirable, to consider political objectives that are technically unfeasible, or that are so expensive that they could exclude some other objectives of the same importance or even more important for the society. Political and technical choices are therefore so interrelated or intermingled in the practical work of the planners and of the decision makers, that it may appear hopeless to try to keep such an apparently abstract distinction.

The main issue, however, is not to separate those elements completely in every practical case. It is rather to understand what are the main arguments to be taken into account in order to reach the "best" solution in very situation where a choice is needed. This is basically a question of rationality: the important fact is precisely that the rules of rationality are different for the two kinds of choices (or if one prefers, for the two types of elements of choices).

On the one side, regarding the economic and technical choices, we are in the realm on instrumental rationality, which commands the selection of the best means adapted to the objectives that have been selected: for example, comparison of two techniques to make an industrial object, to cultivate a cereal, or to map and make a road network; or, more globally, to organize the various expenses from a given budget (of a public collectivity, or of an association) in order to maximize the satisfaction expected from this allocation of resources. On the other side, we can simply consider that there is no "rationality" of goals: there is no rationality, in the strict sense of the word, in the choices of social (and also individual) objectives: these choices are the result of the process of revealing preferences, which themselves proceed from judgments of values – which are not debatable by definition, at least in economic terms – rather than from technical calculations. In the traditional approach of micro-economics, these "tastes", or "preferences" are supposed to be given, and not debatable; they are data for the economic calculations to be made for maximizing the satisfactions of the individual actors. In public (or collective) economics, however, things are a little more complicate, because most of the time, the society does not know, at the beginning of the planning process, what kind of society appears more desirable for itself: the social preferences have to be built systematically, through a process of debate between all the decision-makers considered as "legitimate", or, in a democratic system, between the whole population and its authorities. But in all these cases, the most important thing is that the rules of instrumental rationality do not apply, since the choices are, by definition, about objectives and not about means. When a government decides to declare a war against his neighbor country, or when he decides to starve or to kill a social group by systematic deprivation, or to build a luxurious presidential palace, strictly speaking, the arguments to fight those options are not (or not primarily) instrumental: they are political in the sense that these arguments are based on value judgments about the goals of the society; therefore, they can be criticized on ethical and political grounds, and secondarily on the basis of their coherence with other objectives, but they do not depend primarily on a judgment of "rationality".

Is this a purely conceptual and theoretical consideration? I do not believe so, and the reason for considering this distinction is so obvious that I do not want to start a long explanation of it. I shall only argument by coming back to the political nature of the planning exercise nature and to the consequences of this nature on the way the main choices of a strategy – which are sometimes called "macro-decisions" – are made. A good example, by contrast, of the habitual ignorance of this political characteristics is the role of the international system, and more practically the role of international organizations in trying to influence, or even to impose, the development strategies in the so-called "developing world". One fact is striking in this context: most of the international reports (World Bank reports, for example) on developing countries are "similar", in the sense that most of the time they do not consider any political options in terms of development objectives, but only economic and technical options in terms of means. In my opinion, the reason of this fact is that most of the choices of finalities and goals are nor considered as important, in the minds of most international experts, because after all, most countries are supposed to consider only the so-called "rational" objective of enrichment, in the narrow sense of the growth of production and income, globally or per habitant. When this reasoning becomes predominant – that is: if we eliminate this difficulty of the choice of the most desirable society and of its criteria -, all strategy becomes, by definition, object of instrumental rationality. As a consequence, international organizations may prove that they are intellectually and technically superior about that kind of decisions: the "best" choices of most development strategies are supposed to obey World Bank's and other organizations' recommendations. In this perspective, it becomes possible to consider that development is not at all a political process but is mainly dependent of instrumental rationality... This seems to me to be a very convincing argument for re-examining more seriously the distinction between political and economic choices.

The Role of the Experts in the Planning Process

Shall we conclude that experts should be kept out of the process of development planning as much as possible? Not at all. I think however that this role has to be more strictly defined, in harmony with the distinction proposed in the preceding section.

The story of development planning in many developing countries, especially in Africa, is the story of a misunderstanding between policy makers and experts: the first think (rightly, in a sense) that they are in a position of sovereignty and try to impose their choices, often unrealistically, to all economic and political actors; the second are sure that they are alone to know exactly what can (and should) be done and what cannot be done, and try also, consequently, to impose their choices. The most frequent result is a breakdown and a complete failure of the planning process, or at least a persistent illusion in the efficacy of planning, as well as of the reality of the development strategy which is supposed to be elaborated through this planning.

None of the partners in this misunderstanding is right. The only solution to get out from such an impasse is to build a permanent dialogue between decision makers and experts: the political authorities proposing development orientations and projects, first in broad terms, the second reacting and showing what are the broad possibilities and alternatives; the authorities making then more detailed choices on this basis, until new "crossroads" and new choices are again needed, and then have to be debated between decision makers and experts continuously and permanently, in order to build more and more detailed strategies and projects which have some chances to be at the same time politically desirable and economically and technically feasible.

Macroeconomics and the Role of the Macroeconomic Framework

How are the preceding observations to be reconciled with the predominant image of planning, which seems to be inevitably macroeconomic?

I have said above that the macroeconomic work is often considered as so fundamental, and as the most scientific, and therefore "serious", part of the preparation of the plans: by experts, who think that they have a monopoly of competence in this field, of course, but also sometimes by political authorities, who think that the best solution for building a strategy is to commit the work of preparation to the most "famous" experts they can find. Using various sorts of more or less sophisticated models, this macroeconomic framework aims at building a synthetic image of the evolution of the economy and of its structure, in the past, in the present, and also in the future as an estimate (based simultaneously, most of the time, on national as well as on international projections), but also, without clear distinction, as an objective of the plan itself (for example when the plan tries to show how to accelerate the growth of GDP, of consumption, of exports, or of the school system, etc.).

It is obvious that these techniques have brought a real progress in the accuracy and in the precision of the planning of development strategies, and therefore that they still will have to be developed in the future, for private as well as for public interest reasons. The problem, however, is in the real meaning and, more importantly, in the use made of these projections and modeling, by the public opinion as well as by many economic actors. By the opinion and economic actors, who are tempted to consider these exercises, not as projections of various possible futures, but as predictions of the most probable future. But also by the experts themselves, who are obliged to build some simplified view of a future that they know very little, and who can imagine, and have the time to build, only a limited number of alternative scenarios for this future: at the end, they are in a situation where they believe themselves in their own projections as predictions, while the real future becomes, most of the time, quite different of what they have thought. It would be too easy to mock this attitude, as the exercise is obviously quite difficult, and this weak result probably inevitable.

But the fact that we all know this weakness is not a reason to abandon this type of exercise: the real need is to use it differently. My suggestion would be to link it more closely to the dialogue proposed above between experts and decision makers: the macroeconomic projections would become more precisely a "framework", a set of alternative possibilities in the evolution of the economy (or of a part of the economy), and therefore as a set of information to prepare the alternative choices inherent in the planning process. In this perspective, the alternative scenarios prepared by the experts would be used to show that "if you do this, you will probably get this result, with such advantages and such costs for this and that social group; on the other hand, if you rather do that, you will get quite different results with such and such characteristics". This approach would be more realistic in general, but it also would take a more realistic account of the difference between political choices and economical choices, allowing for a better debate between decision makers and experts about the possible, alternative contents of the "general interest" and the practical feasibility (in terms of "opportunity costs", especially) of these alternatives.

I know that many planners will keep contending that it is basically what they do; but, as far as I can judge from what I have seen myself, I remain skeptical regarding the use made of these exercises as an instrument for the political and economic choices. And the main reason for such a skepticism is probably the technical complexity of these macroeconomic calculations, and the fact that, as a result, many "users" of these calculations tend to "take" them, or to "consult" them superficially, as an esoteric oracle, or at least as purely technical reports that they cannot assess; consequently, they take them without debating or changing any of their components, but they still feel free to neglect them practically in their own decisions. This is, in my opinion, one of the main reasons of the misunderstanding between experts and decision makers and, as a result, one explanation for the irrelevance, or for the inconsistency, of many plans as a global instrument for policy-making.

The Problem of Consistency Between the Global Orientations of the Planning Process and the Specific Policies

I have mentioned already that one can observe fairly often, even in the best articulate experiences of development planning,[2] a lack of connections and sometimes of coherence, between the "general" and the "specific" or "sectorial" part of the plans. Again, this is a very important issue – because the very justification of a planning process is precisely to be an attempt at a better coherence among all the components of a development strategy -, but it is also a very complex, multidimensional issue that could be studied significantly only by examining some specific experiences chosen as examples, and it is not the place to undertake that kind of study here. I only seize this opportunity to make here two basic remarks on the planning process.

The first is just a link with the preceding paragraphs, especially about the function of experts, in relation with the political choices of the objectives of the development strategies, and about the role of the macroeconomic framework as conceived and used by most of the development plans. If there is no connection between the general declarations of the political authorities about their ambitions and development goals, on the one hand, and the technical work of experts and specialists, in building a macroeconomic framework and in elaborating detailed policies in some fields or sectors of activity, on the other, the lack of consistency between all these "pieces" of planning is probably inevitable. The main challenge (and again, the main reason for planning) is precisely to establish this logical connection between all these components; and the basic pre-requirements for such coherence are:

(a) To agree on the very important principle of a permanent dialogue between political authorities and experts (which means, in fact, between political choices and economic-technical choices), and
(b) To use the macroeconomic instruments to make this coherence more precise in a given time perspective.

This seems obvious for any observer of the planning process from outside, but is rarely put into practice, probably because there is a sort of solution of facility which consist, for political authorities as well as for a large part of the public opinion, to consider that the development process is, after all, be conceived exclusively

[2]The Indian planners have been themselves fairly critical about the consistency of their own approach of planning, as an instrument of coherence for the sectorial or specific policies.

by economists, financial specialists, engineers, and experts from international organizations. This is indeed one of the basic misunderstandings that we should try to eliminate about development strategies.

The second remark is much broader and is not directly linked with procedures: it is a basic question about the global function of planning and about the way of conceiving such an ambitious exercise. The majority of difficulties and problems mentioned above may be interpreted as an excess of ambition in the attempts to master the future of the economies and of the societies. Ambition quite understandable of course, and also quite legitimate when compared to the difficulties and threats expected in this future, for economic, social, ecological and political reasons. It appears today, however, a posteriori, that many of the development planning efforts organized in the decades after the Second World War were over-ambitious, because the capacities of the political and technical instruments used were not in proportion with the size of the major challenges of the societies. It was obvious in the case of the centralized economies and of the USSR-models of planning, with the disastrous results that have been observed. But it is also the case for the majority of planning experiences in developing economies (such as India, according, again, to the assessment made today by a majority of the Indian experts themselves) and for planning experiences in market economies (as in France, the story of rising and decline of which may be compared, in many aspects and despite obvious differences in the context, to the Indian experience).[3]

This is again, of course, a complex issue which cannot be described in a few lines: but one of the main reason was probably the one identified, fairly early, in the discussions in East Europe when was raised the question of the "limits of useful planning" Without entering in this discussion, we can draw the main lessons of the debate, which are still valid today, even if they have to be adapted in the present context (especially if one takes into account the accelerated move towards "globalization" in the world economy and, consequently, the new challenges for organizing the management of the national economies).

Basically, today as yesterday, the problem to be discussed is to identify much more selectively, the main themes, sectors and questions where the effort to take a global view – this view being the main ambition of planning, after all – is really worthwhile, when one takes into account simultaneously the constraints and the opportunities for action in a general interest perspective. Clearly, this imposes to make a basic distinction, among the components of the planning process, between what is an assessment of the "context" of development policies (in the past, in the present, and in the future), and the effective will of bringing some normative changes in the results of this context for the national economy or society. The "context" may be seen as the set of constraints and opportunities that condition the policy perspectives; the changes themselves are the political objectives of the development policies. As a whole, the main lessons of the experiences in development planning are

[3] There were 101 priorities in the French Ninth Plan 1984–1988; is there any priority when everything becomes a priority? In fact, the plan was never implemented as a plan, since the main postulates of the policies include in the plan were changed before the end of 1983.

probably the fact that the possibilities of action are more modest than what was conceived and undertaken in the past.

Therefore, the basic challenges for development planners in the future (if any) will be to make a more precise and more practical distinction between what can be done, realistically, and what is condemned to remain a useless but expensive dream. I insist again, however, on the difference between what I have called the context, on the one hand, and the content of the normative policies, on the other. The reason is that it remains extremely important to keep a view as global as possible in assessing the context, while adopting an approach as selective as possible in designing the policies and actions to change this context and its results in the national economy. The necessity of designing global policies does not mean that the policy makers have to change everything simultaneously; it means that it is not possible to bring any useful change, even modest, in a complex situation, if one doesn't take first a global view of this situation to be changed.

Let's take one of the most striking examples of this necessity: the challenge raised by migrations into industrial countries. It is usually not analyzed in global terms; but it is also a very "hot" question, and at the same time it seems to be without any reasonable solution. The problem might be in the apparent obligation of the decision makers to search for an immediate solution, in view of the pressing political constraints in the short run, without trying to broaden the perspective of the problem, in its content and in its space and time dimension. And indeed, this is precisely the main role that planning should play in the world economy today: but this role is somewhat different from the traditional view of planning, which has been too easily attacked (often for good reasons) by the dominant market ideologies in the recent period.

Conditions for a More Realistic Development Planning in the Future

If one accepts the preceding conceptions, it becomes obvious that a lot of changes should be envisaged in the approach for a new development planning. As I have suggested at the beginning of this chapter, however, this is only a preliminary step towards a collective effort of thinking about development planning in this direction. I cannot pretend, as a consequence, to make even an introductory presentation of the conditions needed for a more realistic planning in the future.

As a consequence, I shall limit myself to two interrelated conditions, which are just examples, but which also seem to be also extremely important for this future, in relation with what had been said in the preceding paragraphs. We have to look at

(a) Some criteria to make a distinction between what should be included in the planning exercise, at the analytical level and then in the normative part of the plan; and
(b) Which social groups and actors could be a socio-political base for the normative part of the plan.

We should add, and develop, the need for a more systematic effort of international cooperation, especially in regional unions like the European Union; but it is difficult to present this difficult (and relatively new) question in a few words here.

Criteria for Selection of Sectors and Issues

In the past habits of development planning, there were at least to ways in selecting the main projects or the main sectors of activity as priorities for the plan. One was the direct political choice made by the political authorities, with or without economic detailed justification, for reasons of "economic evidence" (when everybody was ready to agree on the importance of this choice for the future development perspective of the economy), or for reason of national prestige as imposed by the political authorities unilaterally, or for reasons linked with the electoral promises of one or several political candidates. The other way – on which the literature on economic development was very abundant in the 50s and the 60s – was a rather sophisticated method of calculations made on the basis of studies on interbranches relations within the national economy (with the help, when practically possible, of the then new technique of input-output tables, or Leontieff tables, showing how the expansion or the decline of one branch would provoke the expansion and the decline of a series of other branches as a result of the existing and probable exchanges between these branches; or in a less technical way, with approaches focusing on the "linkages effects" between decisions in various branches, as proposed by the famous book of Albert Hirschman (Hirschman 1958).

The "input-output" technique has been considered for a long time as one – not the only one, however – of the most productive approach for selecting the priority sectors; there were a lot of debates about its adaptation in various contexts. The difficulty of such an approach today, however, comes from the globalization process, and from the corresponding acceleration of across borders exchanges in the world economy. As a result, the linkages effects, with precise or imprecise effects, do not have the same meaning or the national economy anymore, because those linkages are much more international than before: the results (costs or benefits) of a decision in the economy of the country A may primarily affect countries B, C or D, instead of country A.[4] In that sense, globalization is obviously a major objection against the idea of preparing decisions in the only context of the national economy.

As, in addition, the ideology of the market made considerable pressures in favor of market mechanisms rather than planning mechanisms, the techniques of national planning have not made much progress in the recent decades for selecting the sectors of priority. A new effort is needed: the solution is not to abandon the whole process of globalization, nor to abandon the whole process of national planning, but

[4] That was one of the unexpected results of the effort to increase the growth of consumption in France in the early 80s, which resulted, not in the growth of national industrial production, but in the growth of imports.

to make a better use of the relations between those two processes in a development perspective. And this perspective could be simultaneously national and international: this shows again the new difficulties of the calculations of the general interest in a global context and in a long-term period.

I don't know personally the solution for this new issue, of course, and I am afraid that much research and experiments are still needed on this subject. It will certainly require a closer cooperation at the international level, at least in some sectors and for some types of decisions; and this requirement will be an additional difficulty for the renewal of development planning.

The above distinction between "analytic" and "normative" components might be one of the elements to be taken into account in this problem of selecting the sectors and the themes in the planning process: the distinction is useful and we can maintain it, but it is probably simplistic. Given their global approach (and, more practically, the scarcity of their time), the planners should try to identify the most important issues and challenges of the time and the space in which they want to make some actions. To select the most "important" challenges is the product of an analysis, but already of a critical analysis, with implicit or explicit value judgments. In this line, again, the components of the international situation (world economy, world political order, regional threats on peace and security…) are certainly part of the needed analysis, but also the analysis of the structure of national interests: the degree of importance of each of them, at the international as well as at the national level, will be directly linked to the kind of interests that will be part of the plan priorities at the national level. For example, the prospects for financial flows and their structure may be more or less important than the prices of raw materials or of food, depending on the importance given to the investors, the productive sectors, the financial sector, the poorest among the consumers, the rural versus the urban population, and so on. The international and regional level is also more or less important according to in the existing institutional organization at this level: it is obvious that the creation of the World Trade Organization, and much more, the building of the European Union and of euro zone changes completely the field open for national planners. This is obvious for everybody. Less obvious is whether, at the national level, the most important criteria for the analysis of the situation must focus on the prospects for economic growth, or on the structure of this growth, or quite differently on the growing inequalities, or on the misery of some groups in terms of basic needs, or on urban management, etc. All these examples suggest that there is no precise border between "analytic" and "normative" components of the planning process (even from a political point of view, with the meaning proposed above): the content of the analysis proposed by a plan is already the result of a political choice.

If we focus more on the normative elements of the plan, the first lesson to be made from the preliminary observations proposed above is certainly that this normative part of the plan should be strictly limited in its content, if the authorities want to keep the effective responsibility for the implementation of objectives and for the management of the priority sectors in this perspective: a short and very selective plan with a very serious control and monitoring of the plan is certainly preferable to a very ambitious plan that no authority is capable of implement. But then how to identify those priority challenges, at least beyond the first critical analysis of a political nature? Some sectors may be difficult to plan in any case, because they are much more dependent on market mechanisms (urban consumption of non basic products): most planners or policy makers would recognize, today, that market mechanisms remain extremely efficient in resources allocation and could be combined with planning mechanisms, for example in terms of broad objectives, or even by using the price systems or other kinds of incentives. Some other sectors may be considered as important in themselves, as components of the general interest, because they affect an important part of the population (directly, as agriculture in most poor countries or indirectly, as transportation or energy infrastructure, or education for the future). This means that again – as in many other remarks made previously – one of the most serious criterion for selecting some sectors, or some activities, or some issues, in the planning exercise should be to look at their practical impact on various social groups, rather than on statistical aggregates at the macroeconomic level. And this point makes the transition with the second condition that I wanted to underline here briefly.

Social Groups, Interests Groups, Socio-political Pressures

This second question is about the need, in any development planning effort, for an organized socio-political support in favor of the planned policies. This support has to be conceived and organized on the basis of an analysis of the social structure of the country or of the society concerned: socio-economic groups in terms of sectorial activities (farmers, workers, civil servants, merchants,) in terms of geographical distribution (by region, urban-rural, etc.), by level of income, maybe also by religious or ethnic affiliation, etc. It is, once more, an obvious requirement as for any kind of policies and political action; this does not mean that this analysis is an easy task, because it this social reality is changing permanently by nature, and because it is meaningful only if this takes into consideration, not only a list of social groups, but also the interrelations between them, their competition or their solidarity, their cohesiveness, etc.

Once more, this condition is so obvious that it should not even be necessary to mention it. Except for one reason: this requirement seems to be, most of the time, completely outside the preoccupations of planners and outside the basic data they use, which focus more on aggregate calculations, on financial flows requirements, and various conditions of macroeconomic consistency. If one accepts the political nature of the planning process, as argued in this chapter, however, it is clear that the analysis of the social reality, in terms of social groups and social interests, as well as the conception of the various policies in terms of costs and benefits for the various elements of the social structure, are a basic necessity for the usefulness of the planning. And this requirement is to be considered from the very beginning of the planning process: if it is the society itself, rather than the experts alone, that is supposed to express its own preferences for the future, the planners have to take into consideration that this society is not an abstraction, it is a collection of social groups, and each of these f groups should be associated to the planning exercise through the permanent dialogue mentioned above.

I would like to go a little further, however, as a consequence of the idea proposed above of a more selective planning, focusing on some critical issues of the development process rather than on an exhaustive approach of the society. If the planning is mainly above some specific issues of the development process (for example: industrial re-structuring in a context of globalization, development of the urbanization network, expansion and diversification of local food production, education or health policies, migration and immigration policies, social insurance building, banking and financial sector re-organization…), then the analysis of the social groups and interests concerned may become less general and much more precise, more specific, and more directly connected with the policy alternative envisaged in the planning exercise. A purely macro-social plan, proposing a sort of grandiose image of the society in the future, may appear as abstract and practically irrelevant as a purely macroeconomic plan conceived in terms of statistical aggregates, as described above.

This more modest approach has to be understood in very practical terms. It is not in contradiction with the need for a political vision that should be present in every planning process: we keep the idea mentioned above of the need for a political vision as a global view of the social reality that the planners want to change, along the preferences of the society rather than in purely technical terms. The crucial point is, once more, the distinction between this global view for analytical and synthetic purposes, and a more specific preparation (in terms of priorities of the plan) of the planned policies and measures: this distinction is probably a useful instrument for planners in order to conceive their role more practically, but also more efficiently.

Conclusions

It may be difficult for the reader to understand where all these considerations above lead us to in terms of economic management, in terms of development strategies, and more precisely, about the future role of planning. There has been a lot of disillusion about planning in the past, in socialist as well as in market economies, at various development levels. In addition, the world has changed, and as a result, the traditional approaches of development planning are certainly less adapted than ever to the challenges of our time. But the recent evolution of the world economy shows also profound disillusion about the system of market economy and of its development performance. Therefore, we clearly are at a turning point of the policy building methods, and we have to imagine other approaches of development strategies, where elements of planning and market will have to be combined differently.

(continued)

There is apparently nowhere, today, a more satisfactory synthesis ready for the present challenges: we need to deepen the social, economic and political thinking on these perspectives. It is in this view that I would like, as a modest conclusion of this chapter about the need for a transition in conceiving planned strategies, to put together some of the major elements mentioned above and to present them, not as a solution, but as a preliminary step for continuing a collective thinking on the future of planning. These elements are especially the following:

- Development planning is primarily an instrument to rethink the future of our societies. This future will be, and has to be, different of the present, because we know that the present situation and the foreseeable evolution are at the same time unacceptable and unsustainable. As a consequence, development planning should remain ambitious and not get lost in technical details. But in order to be efficient, this planning also has to be modest and realistic, focusing on what can be mastered with the present state of our knowledge and of our instruments.
- The obstacles opposing such a conception will inevitably be formidable: because of our ignorance and our lack of experience, of course, but also as a result of the powerful dominant model of development. This model has to be seen as an instrument for maintaining and reinforcing inequalities and privileges of the dominant groups, in all countries, at the expenses of the poor and of the weakest groups, and also for ignoring some of the basic balances in the eco-systems of our planet. Despite the absurdity of some of its ambitions, this dominant model keeps an apparent internal consistency, which will be difficult to change, primarily in political terms, but also in its institutions, with the present combination of market mechanisms and mixed economy characteristics.
- If we want to change the model, we have first to think the changes in terms of finalities and goals – i.e. in terms of the characteristics of the society that we want to promote in the long run -, and not only in terms of technical means (as recommended permanently by the dominant model, with the implicit conviction or belief that all societies pursue basically the same types of objectives as the industrial countries in the world of today). This is the main reason why development planning has to be conceived essentially as a political rather than as a purely technical and economic process.
- As a consequence, for societies who want to be democratic, these goals have to be chosen primarily by the people, by the population, by the citizens, in dialogue with experts but not under their domination. Some instruments are already available (even, paradoxically, among macroeconomic techniques) to organize such a dialogue; they will have to be systematically

(continued)

developed in the future. The three basic functions of such a dialogue are (a) to allow for a broader expression of the various interests of the main social groups composing the society; (b) to study the technical and financial means which can be mobilized for these goals; and (c) to organize a progressive translation of these broad objectives into more concrete and detailed policies and decisions. One major difficulty of this organization in the present world economy, however, will be the globalization of this economy and, as a consequence, the diversity and interrelations between decision makers at various levels, local national, regional and international.

- This is an additional reason to focus the planning process on a limited number of the most important issues, and on a still more limited number of priorities, and not to try to plan everything. This should take the form, probably, of a large debate – in various circles, or at least with social actors who seem to be ready for entering in such a debate – for identifying a limited number of issues where planning is feasible, when planning is conceived as a normative action to be organized in a global perspective.

References

Comeliau C (1999) Planifier le développement: illusion ou réalité? l'Harmattan, Paris

Comeliau C (2007) Is economic growth conducive to development in India? An outsider's reading. http://semioweb.msh-paris.fr/afui/documents/INDIA%20CC%202007.pdf

Hirschman A (1958) Strategy of economic development. Yale University Press, New Haven

Keynes JM (1931 (1928)) Economic possibilities for our grandchildren. Essays in persuasion. Classic House Books

Economic Democracy: Meeting Some Management Challenges: Changing Scenarios in Brazil

Ladislau Dowbor

Abstract Latin America is clearly defining a new path for inclusive and sustainable development. Brazil has been playing a key role in opening the way through the convergence of a set of coherent policies, involving direct transfer for the "fourth world" of critical poverty, intensifying social policies (health, education, culture, housing) at the base of the pyramid, steadily increasing minimum salary, and reducing key environment situations like the destruction of the Amazon forest. The impact is not only politically self-reinforcing, particularly through job expansion, as it is anti-cyclical in terms of the global financial crisis.

Keywords Brazil • Strategic scenarios • Economic democracy • Crisis and opportunities • Pathways to development

In our complex society, limiting democracy to casting a vote every few years, and letting the economy loose in the hands of corporate giants, is simply not working. But the overall governance system that is emerging does not obey our ideological simplifications, such as either planning or markets. What we are seeing in the really existing world economy, is a mix of public planning, market mechanisms, cross-enterprise coordination systems, decentralized participatory management, and the growing and chaotic set of international pacts through which we try fill the gap between a global economy and nation-state governance. Probably nobody knows what kind of animal will result as these different regulation mechanisms interact, but the fact is that we are treading new paths. And simplifications are out.

Ladislau Dowbor is Professor at the Catholic University of São Paulo and consultant to several UN agencies.

L. Dowbor (✉)
Faculdade de Ciências Econômicas, Pontifícia Universidade de São Paulo,
Rua Monte Alegre 984 - Perdizes, São Paulo, SP 05014-901, Brazil
e-mail: ladislau@dowbor.org

F. Mancebo, I. Sachs (eds.), *Transitions to Sustainability*,
DOI 10.1007/978-94-017-9532-6_4

The End of Simplifications

We have called this emerging mix Economic Democracy, which may be overly optimistic, but reflects our view that the economy itself must be democratized, if we want the system to work.[1] The ETH study of the global corporate control network has shown that 737 corporations control 80 % of the corporate world, out of which 147 control 40 %, 75 % of which are financial corporations.[2] An IMF publication, Finance & Development, brings us a pathetic cover-story title: "Who's in charge?" Creating more democratic overall management systems is the key issue. While the "another world is possible" Porto Alegre motto is stimulating, the problem is that another management is necessary for it to happen.

Is Brazil showing the way? This would certainly be an exaggeration, but quite a few obvious innovations have helped the country to start on the long way to reduce abysmal inequality, generate decent jobs, and not necessarily at the cost of destroying our natural resources. In a political environment where if economic and social policies are not pro-rich the elites usually denounce betrayal of democracy, and call in the generals, while the multinationals declare the country non-market-friendly, building more democratic economic management is not easy. What follows is a short description of the main strategic issues.

We are relying here on a wide range of discussions that have been taking place in the *Conselho de Desenvolvimento Econômico e Social* (CDES), the economic and social development council linked to the Presidency, during the last few years, reflecting the broad spectrum of participants and also the many documents, proposals and rulings that have been discussed with the most varied sectors of society, in addition to consultations with experts in the main fields of action. There is a strong convergence overall, notwithstanding the great diversity in proposals. We collected here those, which seemed to contribute most to a systemic coherent outlook, stressing the main lines. We also sought to avoid the temptation of a text that by being so general and prudent would say very little, as it often happens with official papers. This, fortunately, is not an official paper.

In this second decade of the millennium, Brazil is taking off from a new level. In a most impressive way it withstood the worst economic crisis since 1929 and is pointing toward a course essentially based on common sense and a balanced view of economic interests, social needs and environmental requirements. The traditional economic standpoint tied to the simplifications of the Washington Consensus, aged suddenly and is no longer capable of meeting the challenges of a modern and complex society that must look for new expressions of economic, social and environmental policies.

---

[1] The full text of our essay, *Economic Democracy*, can be freely downloaded (Creative Commons) at http://dowbor.org/09economicdemocracykd.doc

[2] An overview of the study published in October 2011 by the Swiss Federal Institute of Technology (ETH), and the link to the original research paper, can be found at http://dowbor.org/2012/01/new-research-on-global-corporate-control-6.html/

Some basic views are gaining ground. For example, that the presence of a strong public sector is not a hindrance but an essential asset, as Brazil's resilience in the face of the financial crisis is showing. Regulation of finance does not imply bureaucratization, it is a safeguard needed against irresponsibility and rampant speculation. To warrant workers better wages and rights do not represent irresponsible and demagogical moves, it represents a straightforward way of generating demand and stimulating the economy. To support the bottom of the pyramid is not charity, it allows for more justice from the ethical point of view, as well as economic common sense since it generates opportunities at the bottom of the pyramid. Investing public resources in the poorer regions, even with temporary deficit, generates new opportunities for future investments through external economies.

Expanding social policies does not mean giving up a greater slice of the economic pie to less productive sectors, it represents stronger investment in people, and this enhances economic development as Amartya Sen has already shown. To support social movements is not to distribute benefits, but to provide working instruments for organizations that have a much deeper knowledge of their economic, social and cultural environment, and are flexible and efficient in their specific domains. Being effectively rooted in the communities is essential for social projects to work, as so many failed "parachute" programs have shown. Stimulating sound environmental policies does not "retard" progress, since energy alternatives, improved family farming and the like generate more jobs and technological innovations than to simply tap on existing natural resources. Maintaining a solid tax basis is not "to take away from the people" as it has been presented, it is an essential instrument for providing more balanced development. Overall, this kind of approach is not based on ideological simplifications, but on a pragmatic approach of expanding policies that have shown to work.

Evidence of improvement does not imply underestimating challenges. The international context continues to be unstable, with a good part of the imbalances of the private financial system in developed countries simply transformed into public deficit, without solving the key issue of the bankers' irresponsibility. In Brazil, social improvements during the last decade notwithstanding, the major challenges remain impressive, requiring more comprehensive initiatives. The whole tax system still awaits greater rationality, fairer distribution of the tax burden, and improved efficiency and redistribution in budget allocation. The modernization of government still depends on rescuing the public dimension of the State, too heavily owned by the corporate world, and the country is still waiting for the increasingly urgent political reform. Environmental policies need to be strengthened and assimilated by the cultures of government and corporations as well as consumer behavior. In some ways, the course to steer has become clearer and society, seeing the obvious results, has become more confident. However, these are early stages of a construction demanding constant rethinking of strategies.

A key point to be considered is the rational use of the country's most impressive potentials and their articulation with new environmental challenges. Brazil has the largest reserve of idle agricultural land on the planet, one of the largest reserves of fresh water and a stable climate, this at a time when pressures for food, feed, fiber and fuels are increasing throughout the world. Brazil masters cutting edge technologies in the biofuels area. The country has an enviable energy matrix, based on hydroelectricity, at a time when mastering the transition to a new paradigm in energy and production technologies is becoming the key for building the future.

In the medium term Brazil will host important international events – the Olympics, the World Cup – that attract even more attention on the global scenario. Increased availability of oil with the Pre-Sal fields opens new perspectives. Adding up these and other factors, if the country can avoid the temptation of one more cycle of agro-exports, or the hasty use of the new oil resources, and is able to protect the environment and to continue improving the new social policy, the virtuous circle enjoys good prospects. Much of the future will depend on how Brazil manages the equation of production, employment, income and environment. Brazil has opened new paths, but the past, and particularly the huge inherited inequality, weigh heavily on the present options.

The slow construction of more performing institutions and more democratic ways of decision, the so-called governance issue, is immensely important. Faced with the political influence of large economic groups and a heavy-handed traditional elite, the government has followed a policy of fragile equilibrium, maintaining privileges of the rich, as a political condition for the development of economic and social inclusion of the poor. Some 150 social and productive inclusion programs have been launched, from the widely known *Bolsa Familia* to less discussed but efficient projects like *Territorio da Cidadania*, *Luz Para Todos*, Prouni, Pronaf and so on.

These programs just work, and they do so because they are negotiated, ensuring a reasonable basis of political support. Furthermore, they also work, in the case of the major social programs, because the first and second tiers of management, people who actually carry the weight of getting results, are generally people who come from social movements and indeed are familiar with the issues, know what type of partnerships must be organized and are knowledgeable about mobilization for the programs. Social movements play a vital role in these processes, and will grow in the future. With all the difficulties in the various sectors, a culture of negotiation, of agreement, of respect towards the interest of diverse segments has been gradually built, however fragile at the onset of this decade.[3]

The outlooks formulated in this text meet certain conceptual definitions considered to be part of the basic set of ideas that are taking shape in the country. As such, first of all, we shall differentiate the concept of economic growth, in a narrow standpoint of boosting gross domestic product, and the concept of development, which involves balanced progress on the economic, social, environmental and cultural levels. The concept of sustainability used here, refers to environmental and social sustainability, in the classic definition of the Brundtland Report, meeting present needs without jeopardizing those of future generations. The concept of local or regional

[3]We presented a study on this new generation of intellectuals in a Latin American Perspectives publication, http://dowbor.org/2011/03/intellectuals-in-a-network-a-new-generation-facing-development-march.html/ or http://lap.sagepub.com/content/early/2010/12/12/0094582X10391066.full.pdf+html

development does not refer to a choice of a particular unit such as the municipality, but to the complex territorial articulations the programs require, and ultimately exert their impact on specific geographic spaces. The concept of planning does not refer to any type of authoritarian central planning, but to consensus building procedures concerning structural programs. This tends to encourage government to build a systemic view of development beyond sectorial reductions, and a long-term view that reduces discontinuity between cycles of government elections. The concept of governance is used here in the broad meaning of management involving the government itself as well as the set of organized social actors involved in decision making.

Unquestionably, favorable winds are blowing. A climate of trust is blossoming. Here there are neither winners nor losers. The best image is that of a high tide that floats all boats. Beyond detail of proposals for the country's various sectors, this is the standpoint: a Brazil that is developing with broader participation in the results, in a sustainable way and by means of democratically negotiated decisions.

The New International Context: Risks and Opportunities

The international financial crisis of 2008 signaled a turning point. Major simplifications concerning the dichotomy between state and market, with their ideological weight, gave way to an attitude of common sense, pragmatism of results, a search for balance. Somehow, to innovate in politics has become legitimate again. Today, this innovative way of thinking is essential. Internationally, the crisis does not disappear. A global GDP of 60 trillion dollars and 600 trillion dollars of global outstanding derivative volumes can only generate chaos.[4] Speculative private sector deficits were turned into public debt, loss of pensions and unemployment. The cost of saving speculators without penalizing them results in new tensions with those that are being called to pay. New regulation mechanisms are being sought, but not implemented. A stable and balanced horizon is not taking shape for the planet. For Brazil, the diversification of foreign relations, with emphasis on South-South and Latin American integration will continue as a priority.

Financially, Brazil today is at radically different level. With 35 billion dollars in reserves in 2002, the country was at the mercy of speculative attacks. Today, with 350 billion in reserves, creditor and no longer debtor of the IMF – which financially is not essential albeit significant in symbolic terms – commercial diversification and better balance between domestic and foreign markets, the country has become an international benchmark. The way Brazil maneuvered among the pitfalls of the 2008 financial crisis, including multinationals repatriating extensive funds from subsidiaries to save their headquarters, was obvious worldwide as proof that

[4] Bank for International Settlements – Nov. 2011 – Committee on the Global Financial System No. 46 – The macrofinancial implications of alternative configurations for access to central counterparties in OTC derivatives markets – http://www.bis.org/publ/cgfs46.pdf – ISBN 92-9131-895-7 (print) ISBN 92-9197-895-7 (online)

common sense and pragmatism are more profitable than ideological simplifications. This trust generated allows Brazil today to even make demands on incoming capital. Success breeds success.

Commercially, a world population, which increases by 80 million per annum with expanded consumption, further enhanced by the biofuels options, should sustain the trend of strong demand for commodities. Brazil, with the largest world reserve of idle agricultural land and 12 % of world fresh water reserves is bestowed with exceptionally strong assets. However, the issue of international regulation of commodity prices, now more dependent on speculative capital movements than on the balance of supply and demand is bound to come into foreground. As an example, global trade of oil reaches 85 million barrels per day, and daily speculative trading reaches 3,000 million barrels.[5] In this respect, Brazil has a stake in promoting a minimum of international regulatory mechanisms.

In geo-economic terms, the trend is towards a shift from the Atlantic basin to the Pacific, with outstanding advances by China and India that represent 40 % of world population and other countries, very dynamic today, such as South Korea and Vietnam, or simply as strong as Japan. This poses structural challenges for Brazil. It should be remembered here that while the United States carried out the Atlantic-Pacific railroad connection in 1890, South-America does not even have a decent highway link between the two oceans. The shift will facilitate a more integrated infrastructure in Latin America, as well as a better balance of occupation and use of the territory in Brazil, still heavily Atlantic oriented in demographics and economic activities. For us, the West acquires new importance.

Another key feature of the new international context is the growing presence of environmental challenges on the planet. While the international financial crisis has migrated from the banks to the ministries, the reality of climate change, extinction of life in the oceans by industrial overfishing, the destruction of forests (particularly important in Brazil and Indonesia), soil erosion, widespread pollution of rivers, groundwater and seas are a matter for concern that, regardless of the Copenhagen, Cancun and other meetings, require added emphasis on the environmental and social sustainability in both the public and private sectors. Brazil holds an advantageous position in this matter.

The social situation is becoming more critical. With the speculative surge in the area of grains, world hunger went from 900 to 1,020 million people. Due to starvation and other absurd causes ten million children die every year. AIDS has already killed 25 million people. The World Bank estimates that four billion people in the world have no access to what they call "the benefits of globalization." These situations are untenable. The social balance of economic policies is becoming more and

[5] Oil prices (Brent) have varied from 12.72 dollars in 1998 to 97.26 in 2008, with huge differences in between. Attributing this kind of volatility to variations in demand, such as Chinese voracity for energy, misses the point of the key impact of speculation (www.oilmarketreport.org). Agricultural commodities fare no better. No steady development planning can exist with such volatility in key world prices.

more of a central issue on the planet and Brazil has shown the feasibility of policies that balance economic and social objectives.[6]

Politically, while the economy that has become largely globalized, multilateral regulation capacity has been dwindling. The balance of power has also been shifting, with a reduction of the power monopoly by the United States and by developed countries in general. The BRICs have started to occupy an international political position, the G-20 begins to open a space for regular negotiation and Brazil, in particular gained an expressive international presence, largely due to the innovative and well-balanced economic, social and environmental model implemented and that is simply working. An expansion of these policies, whose organizational technology made great strides, should be the hallmark of the coming years and strengthen the international role of the country.

In terms of new international context, Latin American integration is increasingly gaining momentum. In the past this policy was characterized by creating more acronyms than facts, and real integration corresponded basically to links between multinational corporations in the region. Today progress in terms of institutions, financing mechanisms, infrastructure (still fledgling), migration mechanisms, the academia itself, is quite evident. Brazil has a key role to perform because of its specific weight, as well as because of the political innovations developed and of the many things in common in terms of inherited social dramas. Latin America is acquiring an identity.[7]

A final key point stems from technological advances, particularly in the area of information and communication technologies. The role of access to knowledge, lower cost of infrastructure and of individual equipment, spreading of global connectivity, expansion of access to knowledge throughout the planet, emergence of numerous economic activities in the so-called society of knowledge – all these changes are taking place at a much faster pace than was expected. Where in the past century major political clashes were over ownership of production means, in the era of the new economy access to knowledge and definition of the new legal framework have become central issues. In Brazil's case, universal access to the knowledge economy presents a new generation of opportunities for productive inclusion and improved quality of life. The challenge is to bridge the gap between technological challenges and domestic educational backwardness.

Overall, on the international scene, Brazil has today a strong role as a key partner, not merely on the basis of its economic strength and cultural wealth but also of practical and common sense proposals in dealing with major social and environmental challenges, as well with the necessary solidarity with struggling countries. The reliability and respect conquered, not only expand the country's leeway, but are

---

[6] Ignacy Sachs, Carlos Lopes and Ladislau Dowbor – *Crises and Opportunities in times of change – 2010,* http://dowbor.org/2010/01/crises-and-opportunities-in-changing-times-jan.html/

[7] An important ECLAC report, *La Hora de la Igualdad,* draws the main line of the new consensus being built. The title, A Time for Equality, is very meaningful. Santiago, mayo de 2010, 289 p. Documento síntese com 58 páginas em português: http://bit.ly/bqwYAh Documento completo en español: http://bit.ly/bA9yrl

intensely reflected as noted in the case of approval of the World Cup and the Olympic Games, in the feeling of self-assurance of the overall population. At this stage, the country really takes off from another level.

A Path Based on Common Sense

As the primary strategic issue, Brazil elected to confront inequalities. This strategy in turns strongly centered on mass consumption as an economic engine. At first, this policy faced strong resistance, but multiplying effects were soon perceived in the process. The main challenge faced by Brazil, the economic and social exclusion of almost half its population, came to be seen as an opportunity, and the country found a new horizon in a growing domestic market. The increasing pressure from the bottom of the Brazilian social pyramid for better living conditions, combined with Government determination to encourage change, created a virtuous circle in which the economic, social and environmental aspects found their common ground.

In Brazil, social expenditures have always been presented as costs, a burden on productive sectors. Economic policies have traditionally been based upon the standpoint that greater competitiveness of the enterprise results from cost reduction. Reducing costs by rationalizing use of inputs and by taking advantage of innovation and technology is indeed essential. On the other hand, cost reduction through cheaper labor reduces the consumer market as a whole and tends to have the opposite effect. Shrinking the consumer market reduces the scale of production, and keeps the economy in the so-called "narrow base," producing little, for a few and at high prices.

It must be recalled, that for an individual enterprise fewer social rights and lower wages do reduce their costs, thereby making it even more competitive in the marketplace. However, this policy adopted throughout the economy reduces mass demand and brings about stagnation in general. In practical terms, what makes sense at the microeconomic level thus becomes an obstacle in broader terms at the macroeconomic level. Redistributive policies applied to the whole economy, such as the improvement of the minimum salary in the last few years, affects all companies, generating a larger markets for all, and reducing unit production costs through economies of scale. This in turn allows the expansion of mass consumption, gradually creating a virtuous circle of growth. If sustained for a longer time, this policy fosters production capacity by stimulating investments, which in turn tends to generate more jobs and increased consumption. Simultaneous expansion of demand and production capacity prompts development without surges of inflationary pressures. The growth spiral becomes balanced. As a matter of fact, sectors that stagnate in wages and social rights are frequently also the ones that become accommodated in terms of innovation in general.

This understanding seldom becomes accepted by theoretical explanations alone. However, when this policy is applied, and the results can be seen, as in Brazil today, many people who were opposed and claimed that favoring the poor was political opportunism, the moods tend to change. In fact, the policy works, and is facilitating everyone's business. Up to a certain point, Brazil has found its course by turning the biggest challenge, poverty and lack of purchasing power that accompanies it, into a vector for expansion of the economy in general. For a long time, it has been said in Brazil that we must make the cake grow bigger, and distribute is later. What we now see is that distribution is what makes the cake rise.

Beyond distribution, a second line of change concerns the expansion of social policies in general, involving education, health, vocational training, access to culture and the Internet, more dignified housing. Here too a traditional outlook is being reversed. The theoretical heritage of neoliberal simplifications is of those who produce goods and services, that is to say the private productive sectors, generate wealth. Payment of taxes on the generated product makes social policies sustainable. Thus the corporation generates wealth, while social policies would represent a cost. Therefore, from this standpoint we should maximize interests of producers, the private sector, and reduce the size of the State, the spender. The real situation is different. When an enterprise hires a 25 year old engineer, this graduated person represents a formidable asset, which has cost years of care, training, access to general knowledge, family sacrifice, use of the most diverse public infrastructure, profiting from the overall technological level generated throughout society. Social policies are not costs but investments in people. And in view of the current progress towards an increasingly knowledge-intensive society, investing in people is what yields the most. In fact, the understanding that production processes of goods and services and social policies are like hand and glove in the development dynamics as a whole, one financing the other, all being at the same time cost and product, points towards a balanced outlook of economic dynamics.

A third key element is the environmental policy. The traditional widely disseminated outlook presents requirements of sustainability as limiting growth, an obstacle for investment and employment, resulting in higher entrepreneurial costs. This is simply the case of a mistaken calculation and already widely discussed at international level, by refuting the argument of externality. Carrying out pre-treatment of emissions in the enterprise where the waste is concentrated is much cheaper than to be burdened later by polluted rivers and groundwater, respiratory diseases and loss of quality of life. For the enterprise it actually comes out cheaper to dump waste into the river; however, the cost to society is incomparably higher. Cutting down the Amazon rainforest does indeed create jobs for a while, but can only maintain them with ongoing senseless destruction. Stepping up investment in sanitation, in turn, creates jobs, reduces healthcare costs and increases systemic productivity. Investing in clean technologies tends to promote sectors that will be more dynamic in the future and improves international competitiveness. To manage our natural resources in a sustainable way, capitalizes the country for future generations, rather than decapitalizing it. Equally important, in the modern global economy a coherent environmental policy generates credibility and respect at domestic and international levels, which in turn opens markets. The truth is that environmental policy has in recent years achieved a different stature and become part of the new economic policy outlined in the country.

A fourth aspect of economic policy relates to reconstruction of the country's capability to plan infrastructure. Good infrastructure, by making access to transport, communications, energy, water and sanitation less expensive, generates external economies for all and enhances the territory's systemic productivity. The cost of freight transport in Brazil is prohibitive, since transporting soybeans and other products of a rather low value to weight ratio, over large distances by truck, generates additional costs for all producers. Rescuing railways, shipyards and coastal navigation, stimulating public transportation in cities, ensuring cheap access to telecom services and broadband, enhancing productivity in the distribution and use of water and especially sewage disposal, the strengthening of renewable energy sources – are initiatives that bring about a huge forward thrust for all economic activities.

Planning and the presence of a solid public administration are essential. Government bashing is shortsighted. Infrastructure provides large networks that interlink the territory. In this sense they are one of the main channels for the reduction of regional imbalances in the country. As an example, expansion in the poorest regions is needed to energize and attract new activities. Public policies can support this type of long-term investment in regions where immediate profits are not realized. This involves planning capacity and a long-term systemic outlook. Brazilian metropolises are coming to a standstill with an excess of individual transportation means and lack of planning. This broader look at the structural needs of the economy is essential for the systemic coherence of infrastructure investments, and should play an essential role in this decade.

Thus, distributive policies rooted in an outlook of social justice and economic rationality, expansion of investment in people by means of focused social policies, gradual assimilation of environmental sustainability in all decision making processes of economic impact, and rational planning of investments in infrastructure that will greatly reduce the Brazilian cost structure by means of external economies – all these trends lead to better quality of life, improved international competitiveness, and gradually shapes a model that, in an environment of democracy and social peace, is opening new paths.

Having a model that not only makes theoretical sense, but that works and convinces many of the economic and social actors in the country is an important asset. None of these policies can be considered new or original. But the fact that through negotiated governance the country has managed to gradually put them together creates a new reality. It also shows how crucial politics can be.

Macroeconomic Policy: Pragmatism and Flexibility

Sound macroeconomic management is also playing a central role. At this point also Brazil is working on a new level. It is a matter of balanced wage policies, prices, credit, foreign exchange, social security, investment and tax collection. Technically complex and subject to constant pressures, macroeconomic policy in Brazil used to follow a neoliberal path that was presented with complex theoretical arguments, but was basically centered on maintaining privilege, and brought about low growth and deeper inequality, always with a semblance of seriousness and austerity. The wage restraint and high interest rates would thus be justified as a means of protecting people against inflation. This area of the economy suffers from an original sin: very few people understand how it works, and therefore it is not subject to democratic scrutiny. And the inflationary past left an imprint on the collective unconscious.

As seen, the overall policy adopted may be summarized as expansion of the economy by a progressive social and economic inclusion, which increases aggregate demand, generating jobs and investment, leading to a virtuous spiral of development. The key element of macroeconomic policy is the balance of the different variables, in terms of amount and timing. The policy adopted was characterized by great flexibility and responsiveness to changing national and international trends, a good dose of pragmatism and the search for balance between interests involved.

In practical terms, the initial phase from 2003 to 2005 was characterized by orthodox macroeconomic adjustments aiming to reassure economic agents that the rules of the game were stable, financial commitments were being met, inflationary pressures were being restrained. In parallel, instruments for management of social policies were being devised, which have as the scarce resource not money but administrative capacity, which develop more slowly. The tax and social security mini-reforms permitted in turn to stabilize accounts. The high prices of commodities and diversification of trade agreements reduced external vulnerability.

The second phase, from 2006 to 2008, is characterized by articulation of policies related to the dynamics of accelerated growth due to inclusion, laying the foundations of current actions. The unified register of poor families – a huge effort to reach 60 million people with no ID, postal address or bank account – the unification of social programs in the *Bolsa Família* (Family Grant), the sharp increase in the minimum wage (therefore also an increase in pensions), expanded support to family agriculture (PRONAF), expansion of credit (payroll loans, financing by BNDES and other state banks), the gradual expansion of investments and other measures led to strong consumption at the base of society and strengthening of private sector investments. The outcome was an impressive expansion of formal employment. In other words, the public administration effectively took over its role of promoting development. Greater demand has not sparked inflation, since the idle productive capacity allowed rapid expansion of supply. Expansion of public expenditures was covered by higher revenues derived from economic growth (over 5 % in 2008) and expansion of the formal economy, allowing the government to simultaneously meet the debt commitments and expand social policies.

The financial crisis of 2008 submitted this policy to a severe test. The extent of the crisis and international panic generated, caused a stall in domestic credit, disruption of private investment, transfer of resources from Brazilian subsidiaries of foreign groups to save headquarters (35 billion dollars in 2008 alone) and an overall climate of insecurity. Faced by falling State revenues, the orthodox standpoint would be to restrain public expenditure with a stringent fiscal adjustment. The government decided on a set of counter-cyclical actions, responding in a rapid and diversified way to the various emerging imbalances. Expansion of the minimum wage was continued (12 % in 2009) generating a positive expectation in the market; critical sectors were stimulated by tax exonerations and incentives; foreign exchange reserves were used to finance exports (foreign funding had dried up completely); public debt financing was reduced to prioritize support for productive activities; state banks were used to stimulate the economy with a broad spectrum of credit lines (for people and business, not for banks); for the lower middle-class sectors income tax rates were subdivided. Instead of being reduced, social programs were increased, and a large housing program, *Minha Casa Minha Vida* (My Home My Life), aiming at one million houses, was launched, thereby generating activities and jobs in a capillary mode for the overall economy.

Bleak forecasts at the time did not materialize. This multifaceted macroeconomic pragmatic policy, based on the understanding that a broader domestic market supports all sectors, simply worked. Even big exporters, like the soybeans and meat producers, found it interesting to be able to compensate the weakness of foreign markets with enhanced domestic consumption. Furthermore, the idea that an active State is needed was endorsed. Today the country continues to face structural challenges, but feels confident in its capacity for macroeconomic management. The private sector feels more secure as to the rules of the game. This decade is indeed starting at a new level.

Regardless of the financial crisis, another vector of economic policy has taken shape and is becoming central, the large infrastructure investments delayed for so long. The Program of Growth Acceleration, the Productive Development Program, expansion of investments by Petrobrás, the PAC II and also the Education Development Plan, plans for widespread access to broadband, the planning of water use and many others are at the same time stimulating investment and maintaining this active scenario. This facilitates all adjustments and introduces in various sectors a structural, systemic outlook, rebuilding planning capacity and long-term strategy definitions. On the other hand, it generates broader pressure on the meager management capacity of the public administration, which had become used to managing privilege rather than promoting development. The country is thus facing new challenges concerning administrative modernization.

If a theoretical outlook should be rescued, it is that macroeconomic balances are dynamic, that it is possible to create demand without excessive inflationary pressure, to increase State initiative without bringing about irresponsible deficit, to find a new balance between domestic and foreign market without exchange dramas, that it is possible to set conditions (presently a 6 % tax) for entry of speculative capital without being declared a "non-market friendly" by the international speculative market and so on. Above all, it is possible to reduce social and regional imbalances without jeopardizing the more affluent sectors and the wealthiest regions, by ensuring that everyone benefits, however the poorer at a faster pace. Common sense works. Just as a high tide floats all the boats, the State may be providential, ensuring that the tide continues favorable.

Conclusions: Groundwork for Further Expansion

Results are now tangible and highly visible. In the words of Nelson Barbosa, a key promoter of these policies, "the facts are screaming." In round numbers the level of formal employment has increased by 14 million since 2002. The formalization generates more revenue, which finances much of the support policy. The minimum wage has increased in real purchasing power by 53.67 % in the period (DIEESE 2010), which affects about 26 million people. Increased minimum wages further enhance the negotiating power of workers. Indirectly favored by this increase are retirees, some 18 million people. The *Bolsa Familia* has reached 12.4 million families, improving the living conditions of around 48 million people. Actually, this means fewer hungry children and certainly less distress in low-income families. Between 2003 and 2010, 29 million people crossed the poverty threshold. Pronaf (family agriculture support) resources were increased from 2.5 billion reais in 2002 to 13 billion in 2009, boosting production of about two million small farmers. The program *Territórios de Cidadania* (Territories of Citizenship), is investing some 20 billion reais in the country's most backward regions through integrated support programs. The *Luz para Todos* (Light for All) program is reaching millions of people now with access to basic domestic equipment. The Prouni, which already has more than half a million students from poor families in universities, also showed impressive results. They do better than others in university, rebutting the argument of down leveling.

The argument that distribution is a kind of unsustainable charity simply does not correspond to reality. Only the *Bolsa Família* is a simple transfer of resources and is relatively a very small portion of the whole. Even so, since it is tied to health and school attendance, it is considered as a social investment. Income at the roots of society leads to immediate consumption of basic consumer goods that improve nutrition, hygiene as well as small family investments that can be verified in each improvement of modest homes, stimulating production of building materials and basic household equipment. The truth is that the multiplier effect of resources is very large when directed to the roots of society. The poor tend to consume local goods, stimulating backward territories. And in terms of quality of life, every dollar available to the poorest families generates an incomparably greater improvement than when it goes to wealthy ones. Social productivity of money falls rapidly as income rises.

The fact is that due to the immense legacy of backwardness, inequality is diminishing in Brazil steadily but still very slowly. The Gini index fell from 0.53 to 0.49. This compares to 0.46 in the United States, 0.33 in Italy and 0.26 in Germany (IPEA 2010). In Brazil all incomes are rising, and faster for the poor than the rich. But, since the starting point is very low for the poor, even a higher percentage here represents small changes in

(continued)

absolute terms. In regional terms, a much faster growth is perceivable in the Northeast and other poorer regions, but here too inequality is falling very slowly (Dowbor 2008).

A central point is that preconceived ideas that effectively fueled opposition against programs aimed at the base of the social ladder are gradually dismantled. Far from "leaning on," the poor are demonstrating an impressively positive ability to use resources. They are poor not because of lack of initiative or creativity, but for lack of opportunity. And indeed the tendency to "lean on" is democratically manifested at various social levels.

Organizing policies for the poorest sector of the population encounters a major hurdle, not lack of resources, but difficulty to manage an extremely capillary support system, for those who often have no mailing address, SSN, bank account or even a birth certificate. In a way, the State did not exist for this 30 % of the population. To carry out the registers, implement communication channels and mechanisms for managing this segment of the population required a huge administrative effort still underway. Thus, an indirect impact of inclusion policies was the implementation of transmission belts between the State machine, local authorities, social movements and ultimately the families. The organizational learning resulting from the *Bolsa Familia*, expanded PRONAF, management committees of the *Territorios de Cidadania* program, numerous national and regional conferences – have all created more effective forms of interaction between the State and society, a vector of better management practices for the future.

In this slow transition to an economically but also socially fair and environmentally sustainable Brazil, progress is undeniable, but social liabilities inherited from centuries of imbalance are large. The country continues with an obvious dramatic inequality. Deforestation in the Amazon has been reduced from 28,000 to 7,000 square kilometers per year, which is a great victory, but it is still a disaster. The metropolitan outskirts are still explosive and require radically larger supportive policies. Backwardness in the quality of education, access to more decent health, generalization of environmental policies, and democratized access to credit are some of the manifold challenges. The country is facing the need to strengthen inclusion policies, as well as the adjustment of the public management capacity and the decision-making processes of society in general. The course to be taken is much clearer today; the basic management instruments are becoming structured. The results that have been already achieved and the experience gained open a new agenda with new challenges.

Looking back, and with the ex-post wisdom we gain from the financial crisis, it is clearly the time to set aside the western version of the Little Red Book curiously called the Washington consensus, the *Tea Party* and *Davos* flag waving so similar in its spirit to the Versailles dancing parties, and the different "silver bullet" solutions throughout the political spectrum. It is a matter of difficult and patient consensus building around the key issues. Economic democracy is certainly not around the corner, but as Ignacy Sachs puts it, in the face of the huge social, economic and environmental issues, we are condemned to innovate.

References

DIEESE – job market in Brazil: recent evolution and challenges, DIEESE, 10 Mar 2010 – http://www.dieese.org.br/ped/mercadoTrabalhoEvolucaoDesafiostexto2010.pdf
For the basic figures, see our Understanding Lula's Strength, 2008 http://dowbor.org/2008/05/understanding-lulas-strength.html/
IPEA – poverty inequality and social policies – 12 Jan 2010, p 9. Statement from the Presidency n. 38 – http://www.ipea.gov.br/default.jsp

Norms, Rules and Sustainable Planning: Who Said What About Norms

Jon Marco Church

Abstract This chapter consists of a literature review of who said what about norms at the international level. It explores what it means to look at sustainable planning as an emerging norm. Given the limited capacity of negotiated agreements to impose sustainable planning as an international norm, the author concludes that it is in the emergence of sustainability as a condition for international agreements where the greatest impact is to be expected.

Keywords International norms • Norm emergence • Sustainable planning • Actors strategies • Governance

What can the discipline of International Relations add to scholarship on sustainable planning? The study of international relations has many limitations. The fact that information is not reliable makes hypotheses hard to test. But it also has some advantages. It forces to think in a holistic manner. I am going to start this chapter on rules, norms and sustainable planning commenting on the anarchical nature of the international system. Then, I will reflect upon what it means for sustainability planning to be considered as a norm in such a context, looking at various contributions made by scholars over the past 40 years or so on norms and related concepts. Finally, I will draw some conclusions related to sustainable planning as an emerging norm at the international level. Planning at the national level does not take place in a vacuum; strong institutions usually support it. Planning at the international level also does not happen in a vacuum either; but it is supported by a different kind of institutions. The international system is anarchical (Bull 1977; Waltz 1979). There is no world government. There are some international organizations with limited powers that serve national governments on the basis of specific mandates. International organizations love programming. There is a cottage industry of action plans and

---

Jon Marco Church is Assistant Professor of Regional Development, Sustainability and Politics at Rheims University (IRCS-IATEUR).

J.M. Church (✉)
IATEUR-IRCS, Rheims University, Campus Croix-Rouge
57 avenue P. Taittinger, 51096 Reims Cedex, France
e-mail: jon-marco.church@univ-reims.fr

F. Mancebo, I. Sachs (eds.), *Transitions to Sustainability*,
DOI 10.1007/978-94-017-9532-6_5

strategic programs, many of which aim at achieving sustainability objectives, such as Agenda 21, Sustainable Development Goals, Europe 2020, etc. Action plans add a strategic component to international relations, which allows focusing resources on priority areas, such as sustainable development.

International plans of action are considered "soft," as opposed to "hard" laws contained in international treaties, agreements and conventions. The latter have a binding character that the former do not have. Plans are not binding, but they normally inform action and may even lead to binding agreements. What does it mean for an international planning tool to be considered a norm? Martha Finnemore and Kathryn Sikkink, who inscribe themselves in the social constructivist current, propose an approach to "international normative dynamics"[1] (Finnemore and Sikkink 1998), which has very much influenced the debate over the past 15 years among international relations scholars. They start from the classical definition of a norm as a "standard of appropriate behavior for actors with a given identity" (p. 891). It is important not to confuse norms from a Political Science perspective, which is close to the concept of institution in Sociology, with the concept of precept (in French: *règle de droit*), i.e. dispositions that are general, impersonal and made compulsory by the state, which derive from the legitimate interpretation of the sources of law. In the form of custom, legal scholars consider norms as one of the sources of law among others, together with legislation, bylaws and jurisprudence. A key element in the theory of Finnemore and Sikkink is the concept of "critical mass" of actors necessary to consider a behavior to be acceptable.

The authors, which adopt a sequential approach, characterize the "vital cycle" of norms in the following manner: we find ourselves in a phase of emergence of a norm until when a tipping point is reached. Beyond this threshold, a "normative cascade" is observed, which leads to the internalization of the principle (p. 895). In this case, the typical example is that of the 1997 Kyoto Protocol: after awareness is raised about the effects of greenhouse gas emissions and the emergence if the principle according to which it is necessary to limit them, a critical mass of UN member countries negotiated, signed and ratified a specific multilateral environmental agreement, which other countries adhered to and that contributed to the internalization of the principle (Depledge 2005; Luterbacher and Sprinz 2001; Oberthür and Ott 1999; Rayner and Malone 1998). From this perspective, this approach distinguishes itself from previous ones because it does not limit itself to the origin of norms, but also looks at successive phases.

If sustainable planning were an emerging norm, the two main independent variables according to Finnemore and Sikkink would be "normative entrepreneurs" and "organizational platforms." Normative entrepreneurs are those actors (members of parliament, civil servants, scholars, citizens, etc.) who "call attention to issues" (Finnemore and Sikkink 1998, p. 897). They would do this based on their interests and following a "logic of appropriateness," which can also be considered as a form of resistance (Lagroye and Offerlé 2010; March and Simon 1958; Olson 1971). Organizational platforms are presented instead as tools (international organizations,

[1]We here consider the terms normative "dynamic," "circulation," and "diffusion" as synonymous.

NGOs, etc.) (Park 2006), which normative entrepreneurs need to "promote their norms." It goes without saying that these organizations "shape the content of norms promoted" (Finnemore and Sikkink 1998, p. 899) and that experts can also constitute platforms along the lines of "epistemic communities" (Haas 1992; Jasanoff and Wynne 1998).

The many studies that follow this approach pay particular attention to the identification of thresholds that determine normative cascades and to elements allowing the internationalization of norms (Finnemore and Sikkink 1998, pp. 905–909). On the basis of existing scientific literature, Finnemore and Sikkink formulate the hypothesis that the need for international legitimation (Bodansky 1999), the prominence of promoting actors (e.g. the diffusion of a "Western" norm is more likely) and the intrinsic features of norms are determining elements. The authors also explore the contexts of norms and identify two more criteria: on the one hand, the proximity of claims to existing norms or contexts where they are recognized (path dependency) and, on the other, world-time, for instance the persistency of a crisis or the concomitance of a meeting under the UN.

An element of the definition of norm that is not sufficiently problematized by Finnemore and Sikkink is certainly identity. Ronald Jepperson, Alexander Wendt and Peter Katzenstein consider values, norms and institutions as the constitutive elements of national identity and, as such, as cultural features (Jepperson et al. 1996). Changing a norm could therefore change or forge a new identity, which could be at odds with the existing identity. However, this is extremely rare. If a norm emerges almost exclusively in often partial, sometimes total opposition to existing norms, it usually does so within a given identity. It is adopted if it is acceptable by most actors or if it is consensual, otherwise it would lose its "normality" and we would find ourselves in a "state of exception," where it is the law of the strongest that is the norm (Agamben 2003; Kuhn 1962; March and Simon 1958; Tournay 2009). In line with classical realism, norms therefore seem to be generally determined by identity and, when it this is not the case, norms become a residual function of power (Aron 1962; Morgenthau 1948); contrary to classical realism, however, different and multiple identities, which are not necessarily reduced to national identity, also seem to have their place in international relations.

Another issue that is absent from the article by Finnemore and Sikkink is a theory of differentiation and of hierarchization of norms. All norms are not equal. Jeffrey Legro proposes an approach to measure the weight of norms (1997). Based on a simple correlation with a measure of "organizational cultures" (practically, a measure of the weight of national norms), he concludes that national norms weight more than their international counterparts (pp. 57–59). Without considering the consequences of this work on the legal debate between monists (there is only one legal order where international norms are superior to national norms) and dualists (the national and international legal systems are separate and the latter possesses a legal character only if it is transposed within national legal systems), it is quite difficult to measure the weight of a norm in an objective manner. In International Law, the only generally accepted distinction is found in article 38 of the Statute of the International Court of Justice based in The Hague: conventions, custom, general principles of law, judicial decisions and teachings of the most highly qualified publicists; on the other hand, the only relatively objective hierarchization is the one between "hard law" and "soft law" (Abbott and Snidal 2000; Karlsson-Vinkhuyzen and Vihma 2009). Less ambitious, Amitav Acharya explores the question of the source of international norms that end up being accepted (2004). While adopting an approach to the identity question similar to the one delineated above, the author looks at the question of localization of norms, i.e. their origin and destination. Are norms originating from Europe more acceptable than norms originating from China? Is it easier to adopt norms originating from the US in Mexico or in the Islamic Republic of Iran? From a perspective that rejects all moral cosmopolitism, this would depend from the "differential ability of local agents to reconstruct the norms to ensure a better fit with prior local norms and the potential of the localized norm to enhance the appeal of some of their prior beliefs and institutions" (p. 239).

Regulation is another concept that helps better understand norm dynamics. It is a concept proposed by the French sociologist Jean-Daniel Reynaud (Reynaud 1989; Terssac 2003). According to this author, there would be a general tendency to the codification of standards of behavior (which may seem bizarre at a time of deregulating globalization). This process would be the object of negotiations among the agents concerned by the social game. Rules are therefore produced by society, while "regulating" it at the same time. What initially looks like a tautology is nevertheless a common perspective about norms within society, which leads to a circular or spiraled model of normation. However, the theory of social regulation of Reynaud does not say much about norms *stricto sensu* (what is the difference and the connection between rules and norms?), about their criteria of acceptability (why norms are "normal?" why do we accept them?) and about the link between norms and identity. Moreover, this theory has limited capacity to anticipate future developments as, similarly to game theory; it is impossible to determine exactly the preferences of agents during negotiations (Schelling 1960, ch. 6).

In 2000, a special issue of the scholarly journal *International Organization* was dedicated to the concept of legalization, i.e. a "particular form of institutionalization characterized by three components: obligation, precision, and delegation" (Abbott et al. 2000, p. 401). Legalization is a process that leads to the codification of international relations along normative lines, where norms are conceived as precepts. Nevertheless, several scholars criticized this perspective as it represents a "narrow conception of law, rooted in positivism, formalism, and western tradition" (Finnemore and Toope 2001; Karlsson-Vinkhuyzen and Vihma 2009). What is certainly interesting in this approach is the idea of a continuum between law and politics and the attempt to decode the process of translation of political relations in legal norms. The editors of the issue did not formulate the hypothesis that this is a general tendency, but that legalization—"hard" and "soft" alike—represents a qualitative difference with important consequences for international relations. Paraphrasing the famous statement by Clausewitz, "law is—according to the editors—a continuation of political intercourse, with the addition of other means" (Abbott et al. 2000, p. 419).

Another interesting concept is that of normalization or standardization (Borraz 2005; Dudouet et al. 2006). The norms referred to in this scholarly literature are not "interiorized rules that are followed without necessarily thinking about them," but a "document containing technical specifications" (Borraz 2005, p. 124, translation mine). The advantage of this approach is of highlighting how "the universe of international normalization is itself victim of a deep antagonism, opposing the supporters of a socialization of international norms to those supporting the globalization of commercial rules" (Dudouet et al. 2006, p. 377, translation mine). On the one hand, we would therefore observe a process of spontaneous socialization of norms while, on the other; we would suffer the violent imposition of norms dictated by economic powerhouses. From this perspective, both the capitalist system and state structures would be the basis of the normalization process. Michel Foucault proposes a historical reading of the "normalization society" and makes a critical distinction between the "norm of discipline" and the "norm of regulation" (Foucault 1976, 1997; Paltrinieri 2010, pp. 62–66). First of all, discipline creates the "normal man." Then, this normality would be considered as natural. Throughout modern era, there would be a transition at the individual level from "human nature" to the "normal man" in the framework of the process that Foucault calls "normation." At the population level, a "normalization" process became visible in the eighteenth century, i.e. the identification of "abnormal" distributions of population. This would happen through three devices: mastering randomness in science, statistical quantification of human multiplicity and the regulation of the distribution of population on its territory (Paltrinieri 2010, pp. 64–66). The latter is a concept that we also find in Physiology, in Biology, in Demography and, we can add, in the practice of Regional Planning.

From another historically charged perspective, normalization was initially developed to make communication objects compatible one with the other (Dudouet et al. 2006, p. 367, translation mine). The existence of a language that allows human beings to understand each other makes us postulate the existence of common elements that allow us to code messages in a manner that others can decode. The nature and origin of these common elements has been object of speculation at least since the time of Plato and Aristotle. Nowadays, we can identify three main perspectives in this debate: on the one hand, those who, with Noam Chomsky, consider language as a mental, often innate faculty (1957); on the other, those who see in language a social and arbitrary convention, such as Ferdinand de Saussure et al. (1988, first published in 1916); finally, there are those who adopt a pragmatic approach to language, seeing in it a communication tool (Eco and Bouzaher 1988). Given that norm is a standard that allows acceptable understanding and coexistence of actors, we could apply to it, by analogy, linguistic approaches: norms could be innate (correspond to "universal values"), conventional (derive from more or less consensual agreements) or simple communication tools. We could then share the communicational approach to norm and law developed by the German philosopher Jürgen Habermas: the latter would result from the tension between facts (things as they *are*) and norms (things as they *should* be) within a communication logic that he opposes to strategic and instrumental reason (Habermas 1997). It is within this logic that legitimacy is found.

Normalization and the communicational logic bring us, in this overview of who said what about norms, to the ancient concept of legitimacy, i.e. "a quality that leads people (or states) to accept authority (…) because of a general sense that the authority is justified" (Bodansky 1999, p. 600; Karlsson-Vinkhuyzen and Vihma 2009; Woodward 2005). According to Bodansky, legitimacy in international relations could derive not only from democracy, public participation, expertise, independence (the European Central Bank, for instance) or procedural equity (transparence, public access, etc.), but also from traditional sources, such as national interest, reciprocity or consent (1999). The concept of legitimacy reminds of the authority emanating the norm and connects its existence and effectiveness to the acceptability of authority itself. According to Max Weber, there are three kinds of power legitimacy or domination: traditional (usage, costume, etc.), charismatic (the extraordinary qualities of a leader) and legal (laws, rules, etc.) (Weber 1995, first published in 1921). Without a world government, this authority is not necessarily hierarchical; at the international level, it is necessarily diffused. Authority would then be subordinated to a "general feeling" of justice, to "fundamental norms" that may be considered immanent (determined by circumstantial elements such as balance of power or randomness) or, in a *regression* ad infinitum, transcendental (derived from universal principles, such as the so-called biblical "golden rule").

If sustainable planning were to be considered an emerging norm, the literature review above suggests several perspectives to study it. I also contributed to this literature by looking at the specificity of international environmental norms, particularly in the case of regional environmental agreements. I found a striking resemblance with non-environmental international regimes in decision-making, with national governments playing a greater role than expected (Church 2011, 2014). This allows us to look at the question of fit and scale from a different perspective (Young 2002, 2010) and explore the role of national governments as ordering factors vis-à-vis an overly fragmented environmental governance (Zelli and Van Asselt 2013).

International norms are often the outcome of negotiations mostly among the governments themselves. At the same time, they structure the collective action of governments (Giddens 1984). As far as international negotiations are concerned, they often follow a logic of acceptability (Olson 1971) and, because of the principle of sovereign equality, it is extremely unlikely to produce international agreements that challenge the status quo, except for tit-for-tat compromises (Axelrod 1984), which can have perverse effects, and win-win solutions, which are rare. Despite small breakthroughs, trying to plan a transition to sustainability through international conventions and action plans will most likely lead to the perpetuation of business as usual, as it is illustrated by the state of play of climate talks.

On the other hand, as we have seen above, norms and the normation process are also a component of international agreements and plans of action (Finnemore and Sikkink 1998; Park 2006). For some reason, a certain idea, a certain behavior is shared, repeated. This is how general international law emerges. It can be resisted, there can even be persistent objectors, such as in the case of the law of the sea, but at some point it becomes normal and the behavior, the idea is taken for granted. Sustainability is somehow already an international norm. It is the product of material circumstances, such as natural hazards and international crises, but also of norm entrepreneurs, such as many participants in the *Rencontres Internationales de Reims*. The normation process can be very similar to regulation. There is, however, an intersubjective component in norms, which often makes norms and principles the condition for acceptability, not the outcome of negotiations. That is, in the case of sustainability, if there is no sustainability component in an agreement, the agreement will not be acceptable. The need for environmental impact assessments, for instance, has become so deeply ingrained that it would be almost inconceivable to agree on large infrastructure in international contexts without them.

Sustainability is clearly an emerging international norm and is becoming a criterion for the acceptability of international projects. Conversely, because of the failure of planned economies, planning has been generally shunned for many decades, even if programming is widely practiced. Several authors in this book reminded us of the centrality of planning for sustainability. The objective of this contribution was not to determine whether sustainable planning is a norm, but to present what are the theoretical perspectives available if sustainable planning was a norm. To any "norm entrepreneur" for sustainable planning, let me conclude by highlighting the importance of mainstreaming sustainable planning in international processes, of getting it "under the skin," of making it a norm. If sustainable planning becomes a criterion for the acceptability of international agreements, the arsenal of instruments available for the transition to sustainability will benefit from a powerful weapon. This could happen spontaneously, but "norm entrepreneurs" can also encourage it.

References

Abbott KW, Snidal D (2000) Hard and soft law in international governance. Int Organ 54(3):421–456
Abbott KW, Keohane RO, Moravcsik A, Slaughter A-M, Snidal D (2000) The concept of legalization. Int Organ 54(3):401–419
Acharya A (2004) How ideas spread: whose norms matter? Norm localization and institutional change in Asian regionalism. Int Organ 58(2):239–275
Agamben G (2003) Etat d'exception. Le Seuil, Paris
Aron R (1962) Paix et guerre entre les nations. Calmann-Lévy, Paris
Axelrod R (1984) The evolution of cooperation. Basic Books, New York
Bodansky D (1999) The legitimacy of international governance: a coming challenge for international environmental law. Am J Int Law 93(3):596–624
Borraz O (2005) Les normes: instruments dépolitisés de l'action publique. In: Lascoumes P, Le Galès P (eds) Gouverner par les instruments. Presses de Sciences Po, Paris, pp 123–161
Bull H (1977) The anarchical society: a study of order in world politics. Palgrave, Basingstoke
Chomsky N (1957) Syntactic structures. Mouton, The Hague
Church JM (2011) La Convention alpine, une organisation internationale: la pyramide à l'envers et le retour de l'état. PhD thesis, Université de Paris 1 – Panthéon-Sorbonne, Paris
Church JM (2014) Ecoregionalism: analyzing regional environmental processes and agreements. Earthscan, London
Depledge J (2005) The organization of global negotiations: constructing the climate change regime. Earthscan, London
Dudouet F-X, Mercier D, Vion A (2006) Politiques internationales de normalisation: quelques jalons pour la recherche empirique. Revue Française de Science Politique 56(3):367–392
Eco U, Bouzaher M (1988) Sémiotique et philosophie du langage. Presses Universitaires de France, Paris

Finnemore M, Sikkink K (1998) International norm dynamics and political change. Int Organ 52(4):887–917
Finnemore M, Toope SJ (2001) Alternatives to “legalization”: richer views of law and politics. Int Organ 55(3):743–758
Foucault M (1976) Histoire de la sexualité (Vol. 1, La volonté de savoir). Gallimard, Paris
Foucault M (1997) Il faut défendre la société: cours au Collège de France (1975–1976). Gallimard-Seuil, Paris
Giddens A (1984) The constitution of society: outline of the theory of structuration. Polity Press, Cambridge
Haas PM (1992) Introduction: epistemic communities and international policy coordination. Int Organ 46(1)
Habermas J (1997) Droit et démocratie: entre faits et normes. Gallimard, Paris
Jasanoff S, Wynne B (1998) Science and decision-making. In: Rayner S, Malone EL (eds) Human choice and climate change: the societal framework, vol 1. Battelle, Columbus, pp 1–87
Jepperson RL, Wendt A, Katzenstein PJ (1996) Norms, identity and culture in national security. In: Katzenstein PJ (ed) The culture of national security: norms and identity in world politics. Columbia University Press, New York, pp 33–77
Karlsson-Vinkhuyzen SI, Vihma A (2009) Comparing the legitimacy and effectiveness of global hard and soft law: an analytical framework. Regul Gov 3(4):400–420
Kuhn TS (1962) The structure of scientific revolutions. University of Chicago Press, Chicago
Lagroye J, Offerlé M (2010) Sociologie de l’institution. Belin, Paris
Legro JW (1997) Which norms matter? Revisiting the failure of internationalism. Int Organ 51(1):31–63
Luterbacher U, Sprinz DF (2001) International relations and global climate change. MIT Press, Cambridge
March JG, Simon HA (1958) Organizations. Wiley, New York
Morgenthau HJ (1948) Politics among nations: the struggle for power and peace. Alfred A Knopf, New York
Oberthür S, Ott HE (1999) The Kyoto protocol: international climate policy for the 21st century. Springer, Berlin
Olson M (1971) The logic of collective action: public goods and the theory of groups. Harvard University Press, Cambridge
Paltrinieri L (2010) Gouverner le choix procréatif: biopolitique, libéralisme, normalisation. Cult Et Conflits 78:55–79
Park S (2006) Theorizing norm diffusion within international organizations. Int Polit 43(3):342–361
Rayner S, Malone EL (1998) Human choice and climate change. Battelle, Columbus
Reynaud J-D (1989) Les règles du jeu: l’action collective et la régulation sociale. Armand Colin, Paris
Saussure F, Bally C, Sechehaye A, De Mauro T (1988) Cours de linguistique générale. Payot, Paris
Schelling TC (1960) The strategy of conflict. Harvard University Press, Cambridge
Terssac G (2003) La théorie de la régulation sociale de Jean-Daniel Reynaud: débats et prolongements. La Découverte, Paris
Tournay V (2009) Vie et mort des agencements sociaux: de l’origine des institutions. Presses Universitaires de France, Paris
Waltz KN (1979) Theory of international politics. Addison-Wesley, Reading
Weber M (1995) Economie et société. Presses Pocket, Paris
Woodward SL (2005) Construire l’état: légitimité internationale contre légitimité nationale? Critique Int 28:139–152. doi:10.3917/crii.028.0139
Young OR (2002) The institutional dimensions of environmental change: fit, interplay, and scale. MIT Press, Cambridge
Young OR (2010) Institutional dynamics: emergent patterns in international environmental governance. MIT Press, Cambridge
Zelli F, Van Asselt H (2013) Introduction: the institutional fragmentation of global environmental governance: causes, consequences, and responses. Glob Environ Polit 13(3):1–13

Part II
Towards a New Social Contract

Rousseau, Rio and the Green Economy

Carlos Lopes

Abstract Rewriting a new social contract implicitly means creating a redistributive system that enhances both intra-generational and inter-generational equity. The Rio rationale 20 years ago is not radically dissimilar to the Rousseauist ideal of freedom and justice. The challenge of a multilateral governance model that advocates sustainable development cannot be severed from one that is able to set global agendas, legitimizes principle of common actions and brings global communities to commit to a process of implementing change at the local, national and international level.

Keywords Social contract • Green economy • Rio+20 summit • Global governance • Equities

Rousseau emerges as the principal source of knowledge for nineteenth century philosophy. It is rare that one man epitomizes such a wide range of attributes – democrat, romantic, educational theorist, botanist, composer, the man who stood for the underdog and the philosopher. In the 1760s, Rousseau's influence on education, sexuality, politics and the self were brought into sharp focus in four of his most compelling literary pieces: The *Social Contract*, *Emile*, *Julie* and The *Confessions*.

The *Social Contract* emerges as Rousseau's most compelling and seminal piece of political theory. It explores legitimate political order in the context of classical republicanism. In his treatise "man is born free but everywhere he is in chains," Rousseau asserts the inalienable rights of the individual and the sovereign "will" of the people. According to Rousseau, freedom is natural, basic and innate. Rousseau's idea of a form of social organization that guarantees social autonomy, and still holds sacred the values of a socially cohesive community, is a recurrent theme in The *Social Contract*.

Carlos Lopes is United Nations Under-Secretary-General and Executive Secretary of United Nations Economic Commission for Africa, based in Addis Ababa, Ethiopia. The views expressed are personal. The author wishes to acknowledge with thanks the support of Rebecca Sabah, and particularly the research support given by Fatima Denton.

C. Lopes (✉)
Economic Commission for Africa, UNECA,
Menelik II Avenue, P.O. Box 3001 Addis Ababa, Ethiopia
e-mail: esoffice@uneca.org

F. Mancebo, I. Sachs (eds.), *Transitions to Sustainability*,
DOI 10.1007/978-94-017-9532-6_6

Rousseau's fundamental belief in collective law remains a timeless principle. In Rousseau's world, equity and freedom are essential lubricants to a functional society. Rousseau's principle of collective governance is kindred in spirit to a multilateral policy system that advocates sustainable development as the principle of governance and institutional infrastructure. Today, 300 years after Rousseau's birth, 20 years after the original Rio Earth Summit, and following decades of multilateral negotiations, Rousseau's principles of social responsibility, civic freedom and collective sovereignty are undergoing sharp scrutiny. In short, Rousseau's well-worn *Social Contract* has unmasked the complexity of re-configuring the world's problems into a singular, dominant global governance regime.

What would Rousseau make of contemporary multilateralist surveillance regime, gridlocked in key areas that have direct links with human security? How would he square with a society that seems to be at odds with the nature/society equilibrium that he staunchly advocated for? Will Rousseau be able to lift today's generation out of the collective myopia that focuses on individualism as the gateway to a prosperous future?

The rise of inequality across the world has revealed new governance challenges and made obvious the shortcomings of the state and market – two critical institutions – to act as regulatory forces. Can the principles of Rousseau's social contract help to square this circle? Three centuries after Rousseau's compelling plea for social autonomy, multilateral institutions have not succeeded in mending the broken pieces of a social contract. Some of the questions that plagued Rousseau's world on inequality, freedom, poverty, nature and society remain relevant in today's society. Nation-states converge and diverge on how to achieve the tenets of sustainable development, the same way Rousseau's ideas divided public opinion of his time; rules remain the basis of social interactions.

We will present arguments for a parallel reading of Rousseau's principles of the social contract in a post 1992 Rio Earth Summit world. If indeed sustainable development were considered as a governance model it would be important to understand what the Rio+20 Summit added to this model. Finally, it will be crucial to examine the perception of asymmetries in today's multilateral regimes and governance.

We shall point to the fact that both Rousseau's principles and those of the two Rio Summits are essentially about change and that both argue for an institutional regime. A regime to, uphold change through rules, social justice and freedom. Institutions, such as the "Sovereign" State, or an international regime such as the United Nations are seen as necessary to chart the course of change. In essence, they determine its contours, and oversee and regulate its enforcement. Rousseau juxtaposes the natural versus the unnatural. He concedes that the maintenance of a social contract is contingent on the process under which members of society determine the social order (Rousseau 2007 (1762), p. 182). This social order is not natural; it is created and maintained by humans in society. When acknowledging the role of social order, Rousseau is also alluding to the complex machinery, processes and sustenance mechanisms that need to co-exist along the vision of social order he advocated for. In today's more complex world, the arguments for a maintenance regime for sustainable development and a fairer society have become compelling.

Rousseau's Social Contract and Sustainable Development: What Parallels Can We Draw?

There are five conceptual arguments that can serve to read Rousseau's contribution to contemporary debates.

First, Rousseau's world of the social contract has several parallels with a post 1992 sustainable development world. As stated earlier, both Rousseau and the iconic Earth Summit are part of change processes. In many ways, the entire concept of sustainable development can be seen as a process of change. However, it is an active process of regulation and self-regulation, adjustments and re-adjustments, with transmutations at all levels.

Second, Rousseau's social contract proposals cannot materialize without some form of association, and an institutional architecture that will devise and uphold the "rules of the game." Equally, a sustainable development regime is maintained by an international structure i.e. a global system. Its enforcement and management will need robust institutions to monitor progress.

"The problem is to find a form of association which will defend and protect with the whole common force the person and goods of each associate, and in which each, while uniting himself with all, may still obey himself alone, and remain as free as before." This is the fundamental problem of which the social contract provides the solution (Rousseau 2007 (1762), pp. 190–191).

Rousseau's essential yardsticks for success resides in the way institutions are sought as a means to maintain social order and cohesion. Rousseau's state of law comes to full representation in an environment of economic institutions. A social contract is borne out of this institutional glue. The existence of a state of law represents institutions in that it describes the rules that determine the manner in which individuals in society deal with each other (North 1990). Rousseau's social contract is strongly equated with good institutions, and can only be sustained if the individuals within the system do not attempt to dislodge it. In Rousseau's view, the state of nature is the natural default action for humankind; yet the danger is that when the state of nature is in place, resources tend to be wasted in expropriation and rent-seeking activities (Cervellati 2005).

Third, Rousseau's narrative of freedom has the same motivations as the notion of sustainable development and the principles embodied in Agenda 21. Agenda 21, the blueprint for how countries can achieve sustainable development gives voice and agency to all stakeholders. It puts development at the center of the debate and local actors as the frontrunners in deciding how strategies can be formulated and actions implemented. Rousseau's freedom narrative may sound ambiguous and even contradictory. In Book I of the *Social Contract*, Chap. 6, Rousseau poses the challenge as he sees it:

> Find a form of association which defends and protects with all common forces the person and goods of each associate, and by means of which each one, while uniting with all, nevertheless obeys only himself and remains as free as before.

The question remains: why must the move to a political society leave everyone as free as before? How does one reconcile the freedom a government has to use coercion to make its citizenry obey its will and the freedom of the coerced citizens? It is clear that the principles of sustainable development cannot be understood in the absence of real freedom.

Sen's idea of expanding the concept of development to include freedom understood as access to basic entitlements is linked to the tenets of sustainability and the Rousseauist ideal. According to Sen, deprivation is strongly associated with the absence of entitlement to "some good rather than the absence of the good itself" (Sen 2009). He argues that in a famine context the default analysis is not an absolute absence of food or poverty, but rather the absence of entitlement to the food that is available. Sen asserts that famine tends not to occur in a country where free press and openness is observed. In short, when victims of famine are able to make visible their plight, governments are compelled to respond. To a large extent, he poses a fundamental question to Rawls and other political theorists such as Rousseau – if justice is reduced to the product of a contract, who will uphold the interest of non-contractors, foreigners and future generations? These interested parties may be overlooked.

The Rousseauist idea of the "general will" is a metaphor for social autonomy. It is indicative of the sustainability of societies acting collectively to ensure that the future generations do not have to bear the burden and correct the wrongs of present generations. The notion of intergenerational equity mirrors Rousseau's "general will" as a symbol of law that will work for the collective good of citizens. Our collective force in a Rousseauist world is when our dependence is de-personalized, and we embrace the community as a way of escaping social ills. The "general will" exercises the main role of reconfiguring forms of dependence. It ensures that society is properly structured to uphold the freedom of each individual. The "General," Rousseau's short hand for the state, will also establish the rule of law to ensure that all members of society are equally treated. Rousseau's sense of "enlightened self-interest," in which individual members of society become recognized, by propping up each other's self esteem, is the same vision of Agenda 21. A vision that reinforces the principle that by acting today in harnessing the Earth's resources, one is merely acting in one's and future generations interests.

Fourth, the notion of power also allows a comparator of Rousseau's "General" to the dominant state and the multiplicity of non-state actors in today's complex world. The management of global problems goes beyond the responsibility and purview of the unitary state actor. This is a very different reality from Rousseau's world where the state was "omnipresent." The implications of managing global issues such as climate change, trade or transboundary resources are not respectful of borders. They tend to "leak" and "spill" over national boundaries (Castree 2003).

The state may exercise its legitimacy and authority within national boundaries, but non-state actors in the form of international regimes continue to assert their authority and governance models, with many countries facing the same global challenges. Today's dominant state, protagonized by the principle of sovereignty, is losing ground. International regimes are in high demand for the expansion of collective territoriality of the state and reduction of transaction costs. They act as providers of information and facilitators of inter-state co-operation (Hasenclever et al. 1997).

With international regimes wielding greater authority in the regulation of global governance processes, the role of the state has been weakened. Rousseau's social contract does not reflect the proliferation of non-state actors in an increasingly complex world. Global challenges such as biodiversity, climate change, and international trade remain state prerogatives. Boundaries confer both sovereignty and exclusivity to the state. When some state powers are shared or ceded to international regimes as part of a process, it is done with a prerogative to roll back any decision contrary to sovereignty interests. As Paterson argues, the "fundamental (yet largely unacknowledged, and certainly unexamined) commitments in this understanding of global environmental politics are of an inter-state understanding of global politics, a liberal understanding of political economy, and of the neutrality of science" (Paterson 2001).

A fifth parallel between Rousseau and the post 1992 world can be found in the immediacy of institutions as emblematic structures for change. It is worth noting that Amartya Sen offers a counter argument to Rawls, and even Rousseau on the importance of institutions as upholding the rule of law. Rousseau's social contract is intimately linked to an institutional order as the main legislator of rules that predetermines social behavior. The naïve assumption is that the right set of institutions will prevail. Little importance is given to contradictory human behavior. As Kant put it: even a "race of devils" could, if intelligent, produce just institutions and a just society" (Kant 1795). The current international governance and decision making processes unmasks this assertion.

Sen's depiction of Sanskrit literature on ethics and jurisprudence outlines the difference between Niti[1] and Nyaya.[2] A careful analysis of both terms reveals their association with justice, but they both summarize different notions. Niti is used to refer to correct procedures, institutions and formal rules; whereas nyaya is a more all-encompassing term that looks to the world that emerges from the institutions we create, rather than merely mirroring the structures of institutions. Hence, Amartya Sen, similar to Adam Smith, North and Mills points to the importance of having a more holistic representation of institutions; looking at them not just through the prism of realization, but, more inclusively, taking into account other factors, such as human behavior.

Understanding Sustainable Development as a Governance Model: Contribution of Rio+20 to the Model

When in 2002, activists, policy makers and stakeholders met in Rio de Janeiro under the auspices of the United Nations, the intent was to chart a course for the future of humanity. This "new" resolve was reminiscent of the commitments that global leaders rehearsed before, pledging then to lift people out of poverty and

[1] Niti – a Sanskrit word in jurisprudence parlance refers to rules and institutions.

[2] Nyaya – refers to law enforcement and regulations.

protect the earth. The Rio+20 Summit was intended as a celebration of the original Earth Summit of 1992. Beyond a celebration of past commitments, Rio+20 was also meant to reaffirm political commitments and help global leaders take concrete actions to move towards a green economy. Twenty years after the iconic Earth Summit, the world had become a more complex place where poverty and inequality remained staple attributes. So, what is the verdict? Well, many pundits describe Rio+20 as a "non-event," a "failure in leadership," a "vague agreement" or a "weak outcome."

Scientists and activists alike had pinned their hopes on a conference that would emphasize the expediency of a world in distress. But, it is not just earth's life system that is under threat; it is more than one billion of people that go to bed hungry every night. This stark reality is certainly an aggravation of what Rousseau observed in the eighteenth century, at least in size and complexity. Rio+20 may appear as a demonstration of how the world is getting worse rather than better.

Critics of Rio+20 seem to have forgotten through the controversy of the 1992 Earth Summit. It was perceived by some critics as having failed to set a new direction for life on earth. When one attempts to fast forward 20 years, one can quickly discern the remarkable positive evaluation the conference received since.

Whatever the complexion of the immediate evaluation of the 2012 UN World Summit on Sustainable Development, it is, nevertheless, clear that countries failed to design their cooperation mechanisms in ways that provide a new momentum for the implementation of Agenda 21. The Summit simply laid bare the fact that global commitments – with strict targets and uniform measurements of progress – were politically unrealistic (Papa and Gleason 2012). Therein lay both the challenge and the paradox.

Three hundred years after the birth of Rousseau and the foundation of social autonomy – can global leaders come up with a "blueprint" to regulate the affairs of so many diverse people, economies, ecosystems and social formations? How can this uniformity in measurement enable and kick start action on key principles associated with the social contract i.e. equity, freedom, the rule of law, etc.?

Yet, kick starting some of the principles of sustainable development has further polarized the world in 2012 Rio, global leaders channeled their energies in defining what green economy is and what it is not. The term achieved diplomatic momentum at the summit. Many developing countries were concerned that this new concept will replace sustainable development. Advocates of sticking just to sustainable development felt that major policy matters on finance and technology were deliberately forgotten in the interest of an even looser term. Hence, the debate was given an ideological and semantic resonance.

Some countries, mostly from the South, asserted that the green economy is simply a component of sustainable development and should not be used to dictate the pace of international policy governance. For richer countries, greening the economy (through clean energy) could be a safe pathway to increasing economic growth and creating new – "green" – jobs. Subsequently, the efforts to adopt a green economy road map with environmental targets, goals and deadlines met with great resistance in Rio.

In addition, some critics argue that Rio+20 was a failed opportunity in its interpretation of "Sustainable Development Goals (SDGs)," the replacement for the UN Millennium Development Goals (MDGs). SDGs were omitted from the General Assembly resolution, which provided the mandate for convening the Rio+20 Summit. However, the SDGs have now regained a new momentum since Rio.

Rio produced the typical asymmetrical relationships, with the EU insisting on emphasis on energy, water resource efficiency, land and ecosystems, as the critical areas for measuring the SDGs, whilst the G77 and China placed more emphasis on greater balance between the three pillars of sustainable development. Another vexing issue is related to the Millennium Development Goals (MDGs) and how these are translated on development agendas across the developing world. Many developing countries are concerned that the high visibility that is given to SDGs might drive the original MDG targets and indicators into obscurity, and would like to see a better manifestation of how the MDGs and SDG will integrate.

Another problem evident in 2012 Rio was the lack of robust institutional arrangements that will champion the implementation of actions decided in Rio in the same way that trade is strongly equated with the World Trade Organization. But, perhaps more controversial is the issues of finance and technology and the means of implementation. The cleavage between developed and developing countries on this topic was even starker. Developing countries argue that leapfrogging environmentally sound technologies should mean structured support from industrialized nations. This was a key plank of the argument of developing countries in 1992 and remains a constant in the negotiations 20 years later. The question of new and additional financial flows and respect for the agreed aid target of 0.7 % has also been avoided systematically.

The overriding question remains: should countries commit to new goals and implementation of new concepts such as the green economy if they are unable to secure pledges made 20 years ago? It seems that Rio+20 did not succeed in answering these questions and failed to chart a clear course that will support many of Rousseau's ideals on social justice and freedom.

Sustainable Development: An Impractical Tool for Global Governance?

Sustainable development was born out of a historical context. The theory was an attempt to resolve the tension between environmental concerns resulting from the ecological consequences of human activities on one hand and economic, social and political concerns on the other. The central tenet of sustainable development resides in the concept of equity and social justice for all. This is often associated with Rawlsian theory that suggests a bias in resource allocation to benefit the least advantaged societies (Rawls 1971). The intergenerational solidarity principle translated into the will that resource management of today should not compromise the well being of future generations, remains popular.

More than two decades after the concept was given visibility by the Brundtland Report, our understanding of sustainable development is still evolving (Newman 2006). Indeed, subsequent international conferences such as the World Summit on Sustainable Development, held in Johannesburg in 2002, reinforced the need for change in the way societies produce and consume as a precondition for achieving sustainable development (ECA 2009). In fact, the Economic Commission for Africa "Sustainable Development Report" emphasizes the importance of moving towards sustainable consumption and production to fulfill the dual aspirations of economic growth and poverty alleviation.

The terminology sustainable development also implies balance, the ability to use the different capitals (social, natural, physical) in ways that do not jeopardize natural support systems (Kates et al. 2001). The amount and distribution of the various capitals matter (Dasgupta 2001; Kates and Dasgupta 2007). The terminology has achieved greater political legitimacy as argued by Brundtland "the "environment" is where we live; and "development" is what we all do in attempting to improve our lot within that abode. The two are inseparable" (United Nations World Commission on Environment and Development 1987).

Yet, in spite of this evolution, sustainable development continues to suffer from definitional vagueness (Happaerts 2012). Most critics of sustainable development tend to see it as far too normative and ambiguous. Incapable of bringing practical solutions to complex development and environmental problems (Newman 2006). To break away from this inherent fuzziness and ambiguity, the term "sustainability" is invariably used as a substitute for the absence of clarity in the path towards development. As Holling argues (Holling 1973), sustainability is the capacity to "create, test and maintain adaptive capability." Development, on the other hand can be a process of environmental management that is evolutionary in nature.

Sustainable Development Model: The "Absence" of a "Blueprint"

The new so called engines of global growth such as Brazil, Russia, India and China have a collective GDP becoming closer to that of Japan, France, United Kingdom, Italy, Germany and the US put together. Yet, in spite of their colossal economies and growth trajectories, their roles in acting as models or champions for sustainable development have been overlooked. What is their potential for achieving sustainable development? China is an example of a country that has achieved growth, but has not necessarily linked growth to the principles of environmental preservation.

Critics argue that the sustainable development concept needs to be more flexible and dynamic, able to lend itself to ecological and social realities. Sustainable development is a process of transformative change – across scales- and governance regimes. Sustainable development requires an enabling environment, robust institutions and a set of rules to be adhered to. These are not processes that one can "stumble" into –will require continuous direction and focus.

Perception of Asymmetries in the Current Multilateral Regimes and Governance

The challenge of a multilateral governance model that advocates sustainable development cannot be severed from one that is able to set global agendas, legitimizes principle of common actions and brings global communities to commit to a process of implementing change at the local, national and international level. This operational space can only happen in architecture with actors that "play" the role of multilateral diplomacy. For instance, the United Nations providing the critical platform for multiparty negotiations is a vehicle for change. It is also the "stage" where forms of multilateral diplomacy can be evaluated, and even contested.

Principles such as "common but differentiated responsibility," "subsidiarity," "the polluter pays," have become synonymous to an institutional structure that is largely perceived as an enforcer. As in Rousseau's social contract, the seeds for a transformative development are deeply rooted in the capacity of the perceived institution and how it induces change.

The asymmetries of the world hitherto anchored mainly on the North/South divide have become even more diffused and stratified, with wide ranging inequalities ranging from technology, science and even to the basic production system. Thus, the expectation that the North will provide the key to unlocking development in the South is a "pipe" dream. Many of the big OECD countries have channeled their energies elsewhere and concerns on how efficient and clean technologies can be transferred have remained rhetorical questions. Global leaders such as the European Union have not succeeded in persuading a disinterested US to take a stronger role in the management of global commons. Consequently, the paradox is that the role of the United Nations in managing the state of equilibrium between the three pillars of sustainable development has become more difficult. The South prevailing viewpoint focuses on environmental degradation as the chief culprit to their growing problems of poverty and deprivation (Najam 2006). The voices of the Group 77 and China seem to become even more discordant than before. Yet we are in a world where coalition politics and key networks increase their bargaining power.

How can Rousseau's social contract principle be given more relevance in a complex world, where present generations are held accountable by future generations? Justice between generations is becoming even more compelling. With growing environmental degradation and economic stagnation, the idea of justice between generations was felt acutely in the 1970s. Indeed, the welfare of future generations has resonated throughout the generations as a predominant ideology, often expressed in "faith in the future." The Renaissance, "rebirth," from sleep and the eighteenth century enlightenment period, all promoted the idea of progress in human affairs. In the nineteenth century world, this continued interest on human progress was associated to the Industrial revolution. However, by the twentieth century the future was mired in pessimism with World War II, the Holocaust and the specter of a nuclear war.

Whatever the strength of this "master narrative" the notion of intergenerational equity and solidarity shape the global governance regimes. One could argue that previous political theorists have not sufficiently thought through the notion of reciprocity. Indeed, the utilitarian principle based on the "greatest good for the greatest number" seemingly placed more emphasis on the quantity of life rather than the quality and how this will put future generations at risk. Rousseau, Kant and Locke present a challenge to the notion of reciprocity. In short, if our current actions have implications for future generations, how can our lives be affected by unborn generations?

Obligations to future generations present a central ethical problem, both in terms of how to approach the reality of an ageing population in most of the developing countries and significant part of Asia and Latin America, and a booming younger population in Africa. Let's consider this conundrum. In the interest of intergenerational equity how can we draw up a new social contract that will take into account changing demographic dynamics?

The answer to this "riddle" will lie in the ability to right the youth asymmetry that the world is currently witnessing. In its United Nations, World Population Prospects (2012), on the global population trends, the United Nations said that the world's population will increase to 7.2 billion and is projected to reach 10.9 billion by 2100. Population growth is likely to increase in the world's poorest countries, with high fertility rates, which are mainly concentrated in Africa. It is estimated that half of population growth between 2013 and 2100 will be concentrated in just eight countries: the Democratic Republic of Congo, Ethiopia, Niger, Nigeria, India, Tanzania, Uganda and the United States.

The current youth dynamics in Africa presents a challenge. It is reported that in less than three generations, 41 % of the world youth will be Africans. It is believed that between 2010 and 2020, African will add an additional 163 million people to its potential labor force. In addition, Africa labor force is set to increase outgrowing China by 2035. Approximately 54 % of Africa's youth is currently unemployed and more than three-quarter live on less than 2 dollar (US$) a day. In Africa, the tendency is that youth unemployment tends to increase with higher education levels. Another constant is that government programs aimed at promoting youth employment tend to be inefficient. This is the case for at least 21 countries in Africa.

This generation of young people has a huge potential to expand Africa's productive work force, promote job creation and entrepreneurship and harness the enormous resources that the continent is endowed with. Poor investment in today and tomorrow's youths can constitute a blessing or a curse for the continent. Balancing the development sheet need to be done in ways that do not leave the majority of the world's population disenfranchised.

But how prepared is Africa to deflect the potential tension that can arise from an urban youth population that is rapidly growing, educated, unemployed, frustrated and lacking a political space? Given the relative stagnation of employment in the 15–24-age bracket, how can Africa design and use a new social contract to ensure that marginalized youth are not written off and become fully absorbed in the economy?

The real challenge of the twenty-first century will be the ability to address this demographic mega trend in a manner that will preserve the interests of future generations. How can a new social contract realign the disenfranchised, the old, the young and the poor back to the center of a development agenda? Today's elderly generation in Europe or Japan is able to enjoy a relatively prosperous old age mainly because their working lives are comparatively more prosperous than those of their parents. To what extent can Europe or Japan sustain its social welfare system without re-negotiating a new contract with Africa's youthfulness?

Conclusions

Rewriting a new social contract implicitly means that there is a level of dissatisfaction with the way our world is configured. How do we create a redistributive system that is "solidaristic" and helps to enhance both intra-generational and inter-generational equity? How do we create new institutions that can lift people out of poverty based on a social contract that seeks to provide security and welfare to the poorest in the remotest outposts of the world?

The Rio rationale 20 years ago is not radically dissimilar to the Rousseauist ideal of freedom and justice and the need for a participatory form of democracy that becomes the model of choice. A wholesale shift from the Rousseauist ideal to a new contract that will take into account intergenerational equity, ensure that institutions are aligned to societal needs will be hard to develop. However, there are real risks for policy makers and humankind in general if we dismiss these ideals as utopian. The collective interest is strongly rooted in the ability to institute the behavioral response that will ensure that, whilst cognizant of a risk sharing approach, opportunities are provided to future generations.

References

Castree N (2003) The geopolitics of nature. In: Agnew J, Mitchell K, Toal G (eds) A companion to political geography. Blackwell, London, pp 423–439

Cervellati M (2005) Hobbes to Rousseau: inequality, institutions and development. Issue 1450 of Discussion Paper, IZA

Dasgupta P (2001) Human well-being and the natural environment. Oxford University Press, Oxford

ECA (2009) Sustainable development report, sustainable development report on Africa 3: sustainable consumption and production for sustainable growth and poverty reduction, United Nations Economic Commission for Africa (UNECA), Addis Ababa, Ethiopia

Happaerts S (2012) Sustainable development and subnational governments: going beyond symbolic politics? Environ Devel 4:2–17

Hasenclever A, Mayer P, Rittberger V (1997) Theories of international regimes. Cambridge University Press, Cambridge

Holling CS (1973) Resilience and stability of ecological systems. Ann Rev Ecol Syst 4:1–23
Kant I (1795) Towards Perpetual Peace. Liberal Arts Press
Kates RW, Dasgupta P (2007) African poverty: a grand challenge for sustainability science. Proc Natl Acad Sci 104(43):16747–16750
Kates RW, Clark WC, Corell R, Hall JM, Jaeger CC, Lowe I, McCarthy JJ, Schellnhuber HJ, Bolin B, Dickson NM, Faucheux S, Gallopin GC, Grubler A, Huntley B, Jager J, Jodha NS, Kasperson RE, Mabogunje A, Matson P, Mooney H, Moore B III, O'Riordan T, Svedlin U (2001) Sustainability science. Science 292:641–642
Najam A, Papa M, Taiyab N (2006) Global environmental governance: a reform agenda. International Institute for Sustainable Development, Canada
Newman L (2006) Change, uncertainty, and futures of sustainable development. Futures 38:633–637
North D (1990) Institutions, institutional change and economic performance, Series political economy of institutions and decisions. Cambridge University Press, Cambridge
Papa M, Gleason N (2012) Major emerging powers in sustainable development diplomacy: assessing their leadership potential. Glob Environ Chang 22:915–924
Paterson M (2001) Understanding global environmental politics: domination, accumulation, resistance. Palgrave, Basingstoke
Rawls J (1971) A theory of justice. Harvard University Press, Cambridge, MA
Rousseau J-J (2007 (1762)) The social contract and discourses. bn Publishing
Sen A (2009) The idea of justice. Harvard University Press, Cambridge, MA
United Nations World Commission on Environment and Development (1987) Our common future. http://www.un-documents.net/our-common-future.pdf, Accessed on 23 Aug 2013
United Nations, World Population Prospects (2012) The 2012 revision, http://esa.un.org/wpp/documentation/pdf/WPP2012.press.briefing_Directors.remarks.pdf. Accessed 23 Aug 2013

Issue Linkage and the Prospects for SDGs' Contribution to Sustainability

Peter M. Haas

Abstract This chapter reviews the dynamics by which policy agendas are assembled at the international level, and then assesses the prospects for creating a meaningful set of Sustainable Development Goals at the United Nations.

Keywords Epistemic community • Sustainable development • Sustainable Development Goals (SDGs) • Millennium Development Goals (MDGs) • Linkage politics

Sustainability and Sustainable Development are now powerful policy frames, which are often used to structure policies and debates at the UN. Yet a paradox stalks Sustainable Development. While there is a political impulse at the UN for promoting Sustainable Development Goals (SDGs), there is not yet a clear consensus about what Sustainable Development, its desirability, is nor how to achieve it. Devising SDGs is difficult, because there is not yet agreement on what constitutes sustainability, what are shared goals at the international level, which issues on the global agenda should be included, nor what are the most effective means for making headway in addressing such issues on their own, much less to effectively capture their interconnections.

This brief essay applies the literature in international relations about assembling agendas for complex issue areas based on elements, which elude consensus to this task of building SDGs.

Peter Haas is a Professor of Political Science at the University of Massachusetts Amherst. His research concerns epistemic communities, global environmental politics, multilevel governance, and the role of science in global politics.

P.M. Haas (✉)
Professor of Political Science, University of Massachusetts Amherst,
Thompson Hall, 200 Hicks Way, Amherst, MA 01003, USA
e-mail: haas@polsci.umass.edu

F. Mancebo, I. Sachs (eds.), *Transitions to Sustainability*,
DOI 10.1007/978-94-017-9532-6_7

SDGs and Sustainability

SDGs are intended to be *norms* to guide the development community from 2015 to 2025 in promoting sustainability. There are multiple types of norms – discursive (Sneddon et al. 2006), generative, aspirational, and internalized – here I refer to aspirational those which states are willing to invoke to create institutional arraignments, be they formal organizations, coordinated policies, resource transfers, and the like. I do not treat norms as those standards by which states are willing to be judged. Sustainability is roughly, an integrative category that links or couples the pursuit of economic, social and environmental activities in ways that at the very least don't interfere with another, and at the most, promote a more just, ecological integrated, equitable future than improves the quality of life. Thus, Sustainability norms are conceptual frames that enable linking projects and policies to promote sustainability. Sustainability, in turn, has two components: substantive measures of performance in a particular domain, and procedural elements relating to increased accountability, improved governance, information flow, and participation. Progress is a measure of movement over time – replacing politically intractable political policy frames with more tractable ones, and also achieving concrete improvements in the pursuit of individual sustainability goals.

The SDGs are intended to replace the Millennium Development Goals (MDGs). The MDGs set eight global development goals for 2015, most of which appear to be on track for being met, although it is dispute to what extent the achievements are due to broader economic trends and to what extent it is due to the intentional efforts by development community to target resources on their behalf.

The MDGs achieved the poverty component because it was based on a shared *norm*, albeit one that had been actively constructed through the focused activities of a global network of development norm entrepreneurs (Fukuda-Parr and Hulme 2011). Poverty reduction, if not elimination, was widely achieved in most areas other than the poorest areas of sub-Saharan Africa. The MDGs on women's rights appear to have been reached, as well as the access to cleaner water, although the climate change ones were not really met, beyond achievements that exceeded the intentional abilities of the major economies. That is, reductions in GHGs largely occurred through market forces based on recessions and fuel switching to natural gas, partially driven by technological change from fracking, partially from responses to high oil policies, rather than energy or sustainability policies.

SDGs are more demanding than MDGs. They require integrating activities, rather than pursuing independent list of goals. Moreover, SDGs require harnessing interconnected activities into a comprehensive effort that will transform modern economies in directions that are more equitable, environmental friendly, and generate jobs.

Research on issue linkage in world politics reveals that progress on coupling interconnected issues into a comprehensive agenda with resources requires agreement amongst states on norms and on the causal understanding about which issues are significantly interconnected (Haas 1980). In addition it requires institutional designs that bring together environmental and economic ministries (or international institutions) in

Table 1 Prospects for comprehensive issue linkage

	States agree on causal linkages	States disagree on causal linkages
States agree on norms	Social learning	
States disagree on norms linkages		Incremental tactical linkage

a manner where both are attentive and neither dominates. Two political processes drive such agenda consolidation: social learning about common causal arguments that identify key threats and their interconnections (Haas 1992), and norm development which constructs the normative or principled commitment to addressing a particular set of issues (Finnemore and Sikkink 1998). Such consensus is primarily at the level of governments and political elite; although the ideas on which political consensus are based often percolate up from domestic and transnational epistemic communities and norm entrepreneurs (Table 1).

We have agreement on neither at the international level. Selectively we have them for some issues, and the best I feel we can do in the short term is identify the topics on which there is agreement, seek to encourage policy learning about their interconnections, and develop governance institutions at the UN that can help promote consensus building on a broader framework and set of interconnected issues. There are pockets of consensus about some of the ingredients, and simultaneous pursuit of these ingredients may yield a more sustainable whole, as well as promoting the prospect of governments and publics learning about the interconnections between global issues, and the need for more sustainable approaches. If we put them on the agenda can move incrementally between them. In the absence of such consensus the best that can occur is traditional logrolling that provides for tactical linkages, which is incapable of formulating a meaningful and resilient agenda that can contribute to sustainability. At best it will yield unbalanced pillars of sustainability. They will be based on short-term possibilities for combining the goals of the most influential parties. The major economic players in the UN currently include the USA, EU, Russia, China, Brazil, Japan, and possibly S. Africa.

What are the prospects for an integrated package? The current UN state of play is not optimistic. The High Level Panel on SDGs – intended to provide the technical consensual foundations – instead tried to combine the rhetorical with the substantive. It has identified 5 possible primary goals, and a longer list of 12 subsidiary goals.

Goals

1. End extreme poverty
2. SD at core
3. Transform economies for jobs and inclusive growth
4. Build peace and effective, open and accountable institutions for all
5. Forge a new global partnership

Subsidiary Goals

1. End Poverty
2. Empower Girls and Women and Achieve Gender Equality
3. Provide Quality Education and Lifelong Learning
4. Ensure Healthy Lives
5. Ensure Food Security and Good Nutrition
6. Achieve Universal Access to Water and Sanitation
7. Secure Sustainable Energy
8. Create Jobs, Sustainable Livelihoods, and Equitable Growth
9. Manage Natural Resource Assets Sustainably
10. Ensure Good Governance and Effective Institutions
11. Ensure Stable and Peaceful Societies
12. Create a Global Enabling Environment and Catalyze Long-Term Finance.

Targets

This is a very clever ingredient in the report, as the MDGs are also believed to have been driven by the targets associated with them. However, the MDGs had the targets drafted after the Millennium Declaration by governments, and were drafted by the secretariat without direct state accountability. It is pretty clear that the UNGA is no longer willing to tolerate such bureaucratic stealth, so the HLP anticipated such political resistance by directly including the targets.

The High Level Panel will still have to reconcile its input with the UNGA, which will take the HDP report into account along with UNSG report in fall when it issues the final political resolution in 12/14.

The UNGA Open Ended WG on SDGs is also working on the issue, and is the political institutions, which will adopt the actual SDGs. But it is largely concerned with the question of the distribution of economic costs for achieving such goals, in particular ensuring that developing countries are compensated by the international community for any additional expenditure on behalf of achieving the goals, and that jobs and technology created by the goals are shared with the developing world.

Ongoing disagreements hinder any normative consensus. Is sustainability a primary goal, or should existing MDGs just be fine-tuned to be made sustainable should the three pillars of sustainability be addressed simultaneously, or should refined targets be developed for each one? While the goals are presumably global, should the targets be global or national, and binding or voluntary? What happens when 192 countries each embrace different national targets? Can we prevent the kind of least common denominator commitments that shaped the Kyoto Protocol GHG emission targets?

There is contestation over what are presumptive sustainability and environmental norms (Iwama 1992; Beyerlin 2007; Sands and Peel 2012). Few of the goals enunciated by the HDP seem to be widely shared norms, in the sense of the UN Charter or those that have been expressed in regular binding commitments and practices, other than ending poverty. Regularly cited normative claimants include the following:

- Precautionary Principle (PP) – in some fisheries conventions, Stockholm POPs, biosafety
- Polluter Pays Principle (PPP)
- Prior Informed Consent (PIC)
- Intergenerational equity
- Common but differentiated responsibility
- Basic human needs
- Human rights – no genocide, civil liberties, rights of children and women.
- Human rights for the environment, modeled on right to water, food, etc.
- Metanorms of multilateralism and sovereignty

There is no normative consensus on environmental protection (Sand 1999). Many incompatible injunctions fill the landscape. Many of the presumptive sources are soft law and not legally binding, and many treaties lack ratification by some major parties. Thus they may suffer from illegitimacy in the sense of commanding neither universal support nor likely to be effective because they do not include all the parties who are capable of resolving the question at hand.

Green Economy, or a transformation to a low carbon economy, is contested, because of concern about jobs and technology transfer, and interim adjustment costs by coal and oil producers.

There is no unified framework for Sustainability; no universal consensus exists about systemic goals, beyond poverty alleviation. We know that progress can be achieved through stealth as well as a direct normative and political assault. Indeed, ecology was once dubbed the subversive science because it focused our attention on these connections.

In the absence of universal and uniform consensus, we should be looking for individual substantive components that seem to contain dual consensus and can be aggregated or amalgamated to systemic sustainability.

From the HLP list of 12, 5 goals are taken from the MDGs, so they are presumably stable (end poverty empower girls and women, education, health, access to water and sanitation). Good governance is included, because institutional infrastructure is always necessary to support sustainability efforts in the UN system. The High Level Political Forum, intended to replace the UNCSD, is the most likely governance institution to carry the weight. It should provide the scientific support for a concerted approach to sustainability. At the very least it should collect sustainability data, assess progress, and organize sustainability assessments. It should also coordinate with existing international science panels, and identify any gaps where further knowledge is required. It should provide for involvement by scientists, private sector and civil society representatives.

Three additional items seem to command widespread support. Sustainable energy is now a common public policy goal that is supported by solid technical analysis (Pacala and Socolow 2004; Knox-Hayes et al. 2013). Food security, despite disagreements about GMOs, is a candidate for the SDGs, as is resource security.

Less clear is that the other two items command universal normative or consensual support: job creation, stable and peaceful societies.

Without mobilizing financial resources behind all of them it will be a hard sell. There won't be enough resources committed to pursuing the goals so that they will fail and lose legitimacy; and the overall SDGs won't be able to command the support of the developing world.

Conclusion
In short, I have considered two strategies for approaching a more sustainable future. One is by harnessing shared norms and causal beliefs behind a direct sustainability agenda, which I have argued is unwarranted. The second is a more piece meal approach, which aggregates agreement on specific elements out of the hope that together they will give rise to second order substantive learning and sustainable transformations. The whole may yet be larger than the sum of its parts, if the parts are chosen to highlight the interconnections between the mosaic pieces that constitute sustainable development.

References

Beyerlin U (2007) Different types of norms in international environmental law. In: Bodansky D, Brunnee J, Hey E (eds) The Oxford handbook of international environmental law. Oxford University Press, Oxford, pp 426–448

Finnemore M, Sikkink K (1998) International norm dynamics and political change. Int Organ 52(4):887–917

Fukuda-Parr S, Hulme D (2011) International norm dynamics and the "end of poverty". Glob Gov 17:17–36

Haas EB (1980) Why collaborate? Issue linkage and international regimes. World Polit 32(3):357–405

Haas P (1992) Introduction: epistemic communities and international policy coordination. Int Organ 46(1):1–37

Iwama T (1992) Emerging principles and rules for the prevention and mitigation of environmental harm. In: Weiss EB (ed) Environmental change and international law. UNU Press, Tokyo

Knox-Hayes J, Brown MA, Sovacool BK, Wang Y (2013) Understanding attitudes toward energy security: results of a cross-national survey. Glob Environ Chang 23:609–622

Pacala S, Socolow R (2004) Stabilization wedges. Science 305(5698):968–972

Sand PH (1999) Transnational environmental law. Kluwer, Dordrecht

Sands P, Peel J (2012) Principles of international environmental law. Cambridge University Press, Cambridge

Sneddon C, Howarth RB, Norgaard RB (2006) Sustainable development in a post-Brundtland world. Ecol Econ 57:253–268

Putting the Individual at the Center of Development: Indicators of Well-Being for a New Social Contract

Arthur Lyon Dahl

Abstract For decades, development has focused on the economy and the national scale of organization, and economic indicators such as GDP have provided measures of progress. Even UNDP's Human Development Index includes GDP/capita and is based on national averages. To show that people are at the center of the post-2015 agenda, measures of environmental, economic and social sustainability need to be balanced with indicators of the advancement and well-being of each individual human being, including material, social, cultural and spiritual dimensions of human progress. Such disaggregation would highlight disadvantaged minorities, gender and class differences, and other priority needs of specific populations. Governments, businesses and civil society organizations could identify how their policies and activities facilitate or hinder human progress at all stages of life. Every member of society and component group could see that development actions are just and equitable in objective and will be motivated to support them and to feel responsible for their implementation. The indicators would provide both measures of legitimacy and tools to evaluate the effectiveness of governance mechanisms. Development success would be measured by the extent to which society maximizes the fulfillment of each individual's human potential at each stage of life.

Keywords Indicators of well-being • Human development • Rio+20 summit • Value-based indicators • Millennium development goals

Development has been a subject of international debate for decades, in particular seeking to balance wealth creation and poverty reduction. It has focused on the economy and on the national scale of organization, and economic indicators such as Gross Domestic Product (GDP) have provided the principal measures of progress.

Arthur Lyon Dahl is President of the International Environment Forum, and a retired Deputy Assistant Executive Director of the United Nations Environment Programme (UNEP). He has been consultant to the World Bank on indicators of development.

A.L. Dahl (✉)
International Environment Forum, 12B Chemin de Maisonneuve,
Chatelaine, CH-1219 Geneva, Switzerland
e-mail: dahla@bluewin.ch

F. Mancebo, I. Sachs (eds.), *Transitions to Sustainability*,
DOI 10.1007/978-94-017-9532-6_8

The overwhelming weight given to economic indicators of development at the national level has frequently been criticized, and efforts to go beyond GDP are now well under way (Stiglitz et al. 2009; Ura et al. 2012a, b). However, these are still indices calculated at the national level, and they often hide significant discrepancies within a country, where a small but wealthy region, elite or economic sector can hide much larger pockets of poverty and deprivation in national statistics. Governments have now acknowledged the need to look for measures of development beyond GDP at the United Nations Conference on Sustainable Development (Rio+20) in June 2012 (UN 2012).

States assembled at the United Nations and other intergovernmental mechanisms find it much easier to adopt lofty goals and declarations for development than to implement them. After many broken promises, it is understandable that the public, and in particular that of the less affluent countries, has good reason to look upon any new commitments with skepticism. The Millennium Development Goals (MDGs) to reduce poverty in the world by 2015 were the first to be specified in numerical terms and to be accompanied by concrete indicators of their implementation (UN 2013). Their partial success, although due mostly to the rapid economic progress in China, provides some hope that the Sustainable Development Goals (SDGs) that are to succeed them might also lead to measurable progress in reducing poverty and steering the planet towards greater sustainability (UN HLP 2013). They could become the basis for a new social contract with the disadvantaged peoples of the world.

One aim of the MDGs was to increase the flow of official development assistance. However the continuing financial crisis and economic difficulties have reduced the capacities of most traditional donor countries to maintain their levels of assistance. It is therefore important that the SDGs take into consideration the efforts of the people themselves, and not just of governments. This requires a broader view of what needs to be measured.

Another problem with the national perspective is that it traps individuals within their country of origin. The present system of nation states imposes great injustices on individuals based on citizenship. An accident of birth determines each individual's nationality, and thus his or her opportunities and constraints for development. Countries generally want to reserve their support to their own citizens, or to others whom they choose to admit as advantageous to their economy and society, and may forcibly eject any others. Immigration is a particularly sensitive issue, and in times of economic difficulties produces a strong xenophobic reaction. Yet globalization has broken down barriers to capital movements, and the World Trade Organization is mandated to remove barriers to trade, while the other logical dimension of globalization, the free movement of people, is not on the political agenda. This is both due to, and helps to maintain, the extreme differences in wealth between nations. Yet in the near future, climate change and resource degradation are expected to displace hundreds of millions of people who will need to find new places to live. Measures of individual development should not be biased by citizenship or immigration status. Every human being has a right to develop her or his potential regardless of where they live.

The absolute priority given to the poor and the protection of future generations was already included in the Brundtland Commission definition of sustainable development (WCED 1987, p. 43). The High-Level Panel of Eminent Persons on the Post-2015 Development Agenda, established by the United Nations Secretary-General after the Rio+20 UN Conference on Sustainable Development, has identified the eradication of extreme poverty and a core focus on sustainable development as the heart of what should be a new social contract (UN HLP 2013). Balancing these two goals represents a difficult challenge, since poverty eradication requires the creation and/or redistribution of wealth and resources, while sustainability requires that human society and its global economy reduce their impact to respect the planet's environmental constraints and potentials. Economy and ecology tend to lead in opposite directions (Dahl 1996). The challenges of climate change, sustainable energy and food security, to mention a few, show that development as pursued over the last several decades has taken us far beyond planetary limits (Rockstrom et al. 2009). One cause has been the very narrow view of development. There are entire areas missing, mainly because the debate has been framed largely in economic and materialistic terms, when we know that "development" to achieve "well-being" is a far more complex undertaking that has important psychological, social, cultural and spiritual dimensions.

The materialist perspective reduces human beings to competitive, insatiable consumers of goods and to objects of manipulation by the market, with an intractable conflict between endless individual consumption and humanity's collective need for equitable access to resources. We desire a world of peace and prosperity, but much of economic and psychological theory depicts human beings as slaves to self-interest. Yet it can be argued that well-being for everyone necessitates a more just and sustainable social order. This would require qualities like moderation, justice, love, reason, sacrifice and service to the common good, which must be harnessed to overcome the traits of ego, greed, apathy and violence, which are often rewarded by the market and political forces driving current patterns of unsustainable consumption and production, in which the well-being of a few is attained at the expense of the many (BIC 2010). A new social contract must have a broader view of human well-being founded on ethical principles.

Human Well-Being

Addressing the concept of well-being requires an exploration at the deepest levels of human nature. Who are we, and what is our purpose in life? For materialists, we are simply a social animal, and our well-being can come from meeting our physical and social needs. Humanists may add an ethical dimension of responsibility for our fellow humans and the environment. For many, the human experience is essentially spiritual in nature, rooted in the inner spiritual reality that we all share in common. Each of these leads to an emphasis on different levels of prosperity and well-being. Assuming that this is an inclusive hierarchy, addressing the highest level should respond to needs at all the others as well.

The ultimate purpose of development should be to improve the prosperity and well-being of each individual on this planet. UNDP has produced the Human Development Index (UNDP 2014) to focus on this in a collective way at the national level, but this hides significant disparities within countries. What is lacking is a way to operationalize the concepts of development to achieve well-being at the level of individuals (Dahl 2012a). It is often weaknesses in individual human capacity to create wealth, innovate, collaborate, manage, and build strong families and communities that are the primary barriers to effective development. Ideally, the best measure of successful development would be that it enables every human being to fulfill his or her potential in life both in cultivating individual qualities, personality and capacities and in contributing to the advancement of society.

In addition, well-being is not a static concept, but is expressed at multiple levels and in different ways throughout a lifetime. It is also experienced as relative both in comparison with others and in relation to the individual's own previous experience. Throughout the human life cycle, individuals develop and achieve well-being in several dimensions, including physical growth and health, security and safety, education, work, financial security, justice and fairness, human rights and freedoms, a place in the community, and cultural and spiritual identity. These dimensions are discussed in more detail below.

To show that people are at the center of the post-2015 social contract, measures of environmental, economic and social sustainability need to be balanced with indicators of the advancement and well-being of each individual human being, including material, social, cultural and spiritual dimensions of human progress. Such disaggregation would highlight disadvantaged minorities, gender and class differences, and other priority needs of specific populations (UN HLP 2013). Governments, businesses and civil society organizations could identify how their policies and activities facilitate or hinder human progress at all stages of life. Every member of society and component group could see that development actions are just and equitable in objective and will be motivated to support them and to feel responsible for their implementation. Such indicators would provide both measures of legitimacy and tools to evaluate the effectiveness of governance mechanisms.

Ethical Basis

People will not support development programs unless they see that they meet their needs and are just and equitable in objectives. Without the commitment of the masses of humanity, implementation is impossible (BIC 1995). Justice must therefore be the foundation of any social contract.

Justice is the first virtue of social institutions, so laws and institutions no matter how efficient and well-arranged must be reformed or abolished if they are unjust. The rights secured by justice cannot be subject to political bargaining or to the calculus of social interest (Rawls 1999). The present economic system, when not controlled by government regulation, is driven by greed and by ends of profitability that often justify unethical means, resulting in growing extremes of wealth and poverty. The sole reliance on economic indicators supports this, while adding social and environmental indicators can serve as a counterbalance.

A second ethical principle for a global social contract is the oneness of humankind. Injustices in the past have often been rationalized by claiming that justice only applied to citizens or to a superior race, class or culture. This is a denial of biological reality. Once you admit that the body of humankind is one and indivisible, you must accept that each member of the human race is born into the world as a trust of the whole (BIC 1995). Each of us thus bears a responsibility for the welfare of all humanity. This collective trusteeship constitutes the moral foundation of human rights, development policy and sustainability. Furthermore, in a world that has globalized, the welfare of each country and community can only be derived from the well-being of the whole planet.

There also must be agreement on the purpose of development. Obviously basic material needs must be met. Beyond this, most people would accept that there is a higher social, even spiritual purpose to life. In general terms, we could say that the real purpose of development is to lay foundations for a new social order that can cultivate the limitless potentialities latent in human consciousness (BIC 1995). It follows that the ultimate function of economic systems should be to provide the peoples and institutions of the world with the means to do this. Wealth creation is necessary, but the goal should be to universally enrich the masses. Society therefore needs new value-based economic models to support a dynamic, just and thriving social order that is strongly altruistic and cooperative in nature, provides meaningful employment for all, and helps to eradicate poverty in the world (BIC 1998). In a world where technology has opened up vast possibilities for development to those that possess it, it is unjust to sacrifice the well-being of the generality of humankind – and even of the planet itself – to the advantages which technological breakthroughs can make available to privileged minorities (BIC 1995).

Dimensions of Individual Well-Being

There are many different ways to look at human development and well-being, from the viewpoints of various academic disciplines (psychology, sociology, education, anthropology, philosophy), or as defined in the many cultures and religious/spiritual traditions of the world. There have been governmental as well as academic efforts to define and measure human development, as well as documents adopted collectively by governments at the United Nations, such as the Universal Declaration of Human Rights, and the Millennium Development Goals. The following cross-comparison of several such sources draws out some widely-accepted dimensions of individual development necessary for well-being.

Human Needs

Psychological research has long identified what Maslow (1943) termed a hierarchy of needs:

- Physiological needs (breathing, food, water, sex, sleep, homeostasis, excretion)
- Safety needs (security of: body, employment, resources, moral certainty, the family, health, property)
- Love and belonging (friendship, family, sexual intimacy)
- Esteem (self-esteem, confidence, achievement, respect of others, respect by others)
- Self-actualization (morality, creativity, spontaneity, problem-solving, lack of prejudice, acceptance of facts).

While these are not necessarily hierarchical or always pursued sequentially, a deficiency at a lower level can interfere with individual development at higher levels. Later commentators have suggested that the relative importance of social needs (esteem) and higher individual needs (self-actualization) will vary between individualistic and more collective cultures. While our understanding of human needs has evolved far beyond Maslow's "pyramid," all its levels need to be reflected in any definition of human development.

Characteristics of Being Human

A similar perspective comes from recognizing four fundamental characteristics of a human being. The first is as a biological organism with purely physical requirements for life. Secondly, as a social organism, a person has emotional or psychological needs that can only be met through relationships with others in a family, community and society. Thirdly, as a thinking and reasoning being, there are intellectual needs and capacities to develop; Maslow himself recognized a desire to know and to understand. Finally, all religions and many cultures would identify a spiritual dimension of life as the highest realization of human purpose, including acquiring spiritual qualities, refining one's character, and contributing to the advancement of civilization. To be inclusive, the measures of human individual development would logically include all these levels.

Millennium Development Goals

The first six Millennium Development Goals (UN 2013) address some of the most fundamental barriers to individual development and well-being:

- Goal 1: Eradicate extreme poverty and hunger
- Goal 2: Achieve universal primary education

- Goal 3: Promote gender equality and empower women
- Goal 4: Reduce child mortality
- Goal 5: Improve maternal health
- Goal 6: Combat HIV/AIDS, malaria and other diseases

The dimensions of basic needs, health and education must obviously be included as necessary requirements for well-being, along with special efforts to ensure the development of the half of the human population that is female. Every life lost to poverty or disease is a complete failure to achieve well-being.

Human Rights

Another way to identify the dimensions that need to be included in individual human development is through the human rights agreed internationally in such instruments as the Universal Declaration of Human Rights (UN 1948). Denial of a human right is a denial of one or more enabling conditions for, or forms of, well-being. On this basis, we should include the dimensions listed in Table 1.

Life-Cycle of Needs

Well-being is not a fixed goal to be reached by each person at some point in time, but is reflected continually in a cyclical process of individual progress or evolution, from infancy and childhood to adulthood, reproduction and finally old age and death. Throughout this process, each individual has qualities and potentials to be discovered and cultivated. The types of development and their priority change at different stages of this life cycle. Where individual progress is initially dependent on others, and the family or some substitutes for it are primordial for healthy human development and well-being, autonomy increases until the individual has almost complete responsibility for setting and achieving objectives in life. The increasing dependence in old age may reduce well-being in a material sense, but there is clear potential for continuing advancement in spiritual virtues like patience and detachment.

The quality of individual development at one stage is often an important determinant of the development possibilities and limitations at subsequent stages. A child physically and mentally stunted by malnutrition, or emotionally impacted by abuse, will have reduced potential for many kinds of future development. Well-being would therefore need to be considered at different critical stages in the human life cycle.

Table 1 Dimensions of individual well-being in the universal declaration of human rights (Article number)

Free and equal in dignity and rights (1) without distinction (2)
Life liberty and security of person (3) no slavery (4) no torture (5) no arbitrary detention (9)
Recognition before the law (6) equal protection, no discrimination (7) effective legal remedy (8) fair and public hearing (10) presumed innocent (11)
Privacy, family, home, correspondence, honor, reputation (12)
Freedom of movement and residence within State, right to leave country and return (13)
Right to asylum from persecution (14)
Right to nationality, and to chance nationality (15)
Marriage and family, protection of family (16)
Right to own property (17)
Freedom of thought, conscience and religion, and to change religion/belief (18)
Freedom of opinion and expression, to receive and impart information and ideas through all media regardless of frontiers (19)
Peaceful assembly and association (20)
Take part in government, to vote, equal access to public services (21)
Social security (22)
Economic, social and cultural rights for dignity and free development of personality (22)
Work, employment, favorable conditions, equal pay for equal work, just and favorable remuneration, protection against unemployment, social protection, form and join trade unions (23)
Rest and leisure, reasonable working hours, holidays with pay (24)
Standard of living, food, clothing, housing, medical care, social services (25)
Security in the event of unemployment, sickness, disability, widowhood, old age or other lack of livelihood (25)
Special care for motherhood and childhood (25)
Education, full development of human personality, understanding and friendship among all groups (26)
Cultural life, arts, scientific advancement (27)
Author's rights to scientific, literary or artistic production (27)
Social and international order to realize these rights (28)
Duties to community for free and full development of personality, respect rights and freedoms of others (29)
Meet the just requirements of morality, public order and the general welfare in a democratic society (29)

Recent Indicators of Happiness and Well-Being

The recognition of the inadequacy of purely economic indicators has led to a number of initiatives to "measure the immeasurable" in terms of human values, well-being and happiness. These are now producing methodologies that make it possible to consider these higher dimensions of human well-being in a quantifiable way.

Bhutan was the first country to assess the purpose of development through Gross National Happiness (Ura 2012a, b; http://www.gnhc.gov.bt/). This is an important attempt to capture a culturally-relevant view of both material and spiritual

Table 2 Domains of Bhutan's gross national happiness

1.	Psychological wellbeing
	Life satisfaction
	Emotional balance (positive and negative emotions)
	Spirituality
2.	Health
	Self-reported health status
	Healthy days
	Long-term disability
	Mental health
3.	Education
	Literacy
	Educational qualification
	Knowledge
	Values
4.	Culture
	Language
	Artisan skills
	Socio-cultural participation
	Driglam Namzha (Way of Harmony: formal etiquette)
5.	Time use
	Working hours
	Sleeping hours
6.	Good governance
	Political participation
	Political freedom
	Service delivery
	Government performance
7.	Community vitality
	Social support
	Community relationships
	Family
	Victim of crime
8.	Ecological diversity and resilience
	Pollution
	Environmental responsibility
	Wildlife
	Urban issues
9.	Living standards
	Household income
	Assets
	Housing quality

development in nine domains with 33 clustered indicators containing 124 variables (Table 2). "In the GNH Index, unlike certain concepts of happiness in current western literature, happiness is itself multidimensional – not measured only by subjective well-being, and not focused narrowly on happiness that begins and ends with oneself and is concerned for and with oneself. The pursuit of happiness is collective, though it can be experienced deeply personally. Different people can be happy in spite of their disparate circumstances and the options for diversity must be wide" (Ura et al. 2012a, p. 1). As the Prime Minister of Bhutan put it "We have now clearly distinguished the 'happiness' … in GNH from the fleeting, pleasurable 'feel good' moods so often associated with that term. We know that true abiding happiness cannot exist while others suffer, and comes only from serving others, living in harmony with nature, and realizing our innate wisdom and the true and brilliant nature of our own minds" (Ura et al. 2012a, p. 7).

Japan has now announced plans to measure national happiness with 132 numerical indicators covering socioeconomic conditions, physical and mental health, and social relations, as well as intergenerational and international differences, and sustainability. The OECD is also developing international standard measures of well-being, and the European Environment Agency is working on a well-being index. Other countries and international organizations are also working on indicators of well-being and happiness.

In 2012, the Earth Institute of Columbia University launched the first *World Happiness Report* at the UN (Helliwell et al. 2012). This has drawn on data from the Gallup World Poll, the World Values Survey, the European Values Survey and the European Social Survey to assess subjective well-being or happiness, both as felt at one point in time (affective) and as evaluated in a reflection on life satisfaction. It demonstrates the validity and policy relevance of such subjective measures, and encourages their widespread use in other surveys. To explain the variations in happiness, it analyzes both external factors (income, work, community, governance, values and religion) and personal factors (mental health, physical health, family, education, gender and age).

Well-being or happiness indicators would represent the summary overall impact of successful individual development. They reflect an integrated perspective that would capture dimensions not shown in the individual characteristics or levels of development.

Values-Based Indicators

A recent research project in Europe on values-based indicators of education for sustainable development (Podger et al. 2013; http://www.esdinds.eu) has developed a variety of indicators for individual values such as empowerment, integrity, justice, trustworthiness, unity in diversity, and respect and care for the environment, that can lead to well-being for an individual and for a group. While these indicators are tools designed for use at the project and organizational level, they also favor behaviors that can strengthen social relationships and increase well-being collectively, as well as at the individual level (Dahl 2012b).

Stages of National Development

While human needs, potentials and desires are reasonably universal; the economic development context within which an individual is born and lives will condition many aspects of both the possibilities for and results of individual development and well-being. These range along a spectrum from a country or region with indigenous populations living traditional lifestyles; to an economically-poor country with rural subsistence farmers, primary and extractive industries and urban slums; through stages of industrialization, either with large multinationals and foreign direct investment, or with many small and medium companies and a few large domestic conglomerates; to a largely tertiary and services-oriented economy. Measurements of individual development need to reflect and be responsive to the individual possibilities at each of these stages of development, and to show how increased development of individual potential helps a country to graduate from one stage to another. How well-being is achieved and perceived will also be very different at each of these stages of development.

Dimensions of Well-Being

Combining all these approaches and extracting a synthesis has produced the following dimensions of human well-being in a more-or-less hierarchical arrangement from physical and environmental through economic and social to the more intangible.

Physical Growth/Health

- Access to basic foodstuffs, food security
- Access to clean water and sanitation
- Adequate standard of living
- Mental and physical health care, access to primary health care, preventive and curative medicine
- Access to energy (cooking, heating, lighting, modern appliances)
- Adequate shelter, housing
- Clean and unpolluted environment
- Possibilities for rest and recreation, physical fitness
- Special care for motherhood and childhood
- Assistance with disabilities and handicaps
- Care for the elderly

Security and Safety

- Life, liberty and security of person
- Protection from slavery, torture, arbitrary detention
- Security of home and family
- Safety from disasters, unsafe conditions, excessive risks of physical harm
- Protection from domestic violence
- Freedom from crime, corruption in everyday life
- Security from military action, violent repression, terrorism

Education

- Literacy, access to knowledge
- Formal, informal and continuing education
- Full development of human personality
- Education to understanding and friendship among all groups
- Work skills, retraining
- Ability to invest in education
- Access to and participation in scientific advancement and technology development
- Access to information and communications technologies

Work

- Right to work, employment, informal sector, subsistence, entrepreneurship opportunities for wealth creation, economic activity
- Just and favorable remuneration, equal pay for equal work
- Ability to meet own needs and provide for family
- Favorable work conditions, protection against unemployment, social protection, freedom of association, time for rest and leisure, reasonable working hours, holidays with pay
- Author's rights to scientific, literary or artistic production
- Access to extension services, technical advice, business management advice, legal advice, accounting services
- Business access to bank account, credit, microcredit, business license
- Effective process for litigation, dispute settlement, legal assistance

Financial Security

- Protection of real value of income, savings, capital and pensions from inflation
- Access to financial services: payments, savings, credit and insurance

- Reliable and adequate money supply, means of exchange, convertibility
- Protection from banking failures, fraud, undisclosed risks
- Security from theft, identity theft, unlawful dispossession, kidnapping, piracy, extortion

Justice and Fairness

- Recognition before the law, equal protection
- Effective legal remedy, fair and public hearing, presumption of innocence
- Low level of income inequality, fair distribution of wealth
- Upward mobility with effort
- Fair taxation, equitable share of responsibility

Human Rights and Freedoms

- Personal freedom and initiative, equality in dignity and rights, free development of personality
- Freedom of speech, right to hold and express opinions, to receive and impart information and ideas through all media regardless of frontiers
- Right to peaceful assembly and association
- Freedom of thought, conscience and religion, and to change religion/belief
- Right to privacy of person, family, home, correspondence
- Protection of reputation
- Right to own property
- Free movement and choice of place of residence
- Right to a nationality, and to change nationality
- Protection from all sorts of discrimination including gender, etc.
- Equal access to public services, right to social security
- Right to take part in government, to vote, to participate in political life

Place in the Community

- Personal status and dignity
- Social networks, friends to count on
- Marriage and family, procreation and raising children, united family circle, protection of family, divorce
- A community respecting public order and morality
- Community trust, reciprocity, resilience
- Participation and empowerment

- Mobility, public transport, access to markets
- Security in the event of incapacity, sickness, disability, widowhood, old age or other unavoidable lack of livelihood
- Old age security (pension etc.)

Cultural and Spiritual Identity

- Right to a cultural identity, heritage and cultural diversity, a sense of belonging (having, retaining cultural roots and knowledge)
- Having a value system, beliefs, ethics and morals
- Vision and purpose in life, hope for a better life, a better world
- Ability to develop the potential in human consciousness
- Participation in culture and the arts
- Access to beauty, to nature
- Overall evaluative well-being or life satisfaction

Indicators of Individual Development and Well-Being

For each of the dimensions of individual development, fulfillment or well-being, it is possible to identify or formulate relevant indicators that assess the presence, absence or quality of each dimension at the individual level. These could show the numbers or percentage of a population with a positive value for the indicator, or could target the laws, institutions and processes designed to assist each individual to fulfill their life in that dimension. Indicators of well-being or happiness can also reflect the result of the development process for each person. Values-based indicators can also be used to assess the more intangible aspects of individual motivation and commitment to sustainable social, environmental and economic development.

Obviously it will not be possible to collect data from everyone, but statistical sampling procedures are sufficiently well developed to determine representative samples of each significant category of a population. Some data can be collected during population censuses, or with questionnaire surveys. With modern communications tools and data mining, and the reuse of data collected for other purposes, measures of many of these dimensions are within reach. After some initial intensive efforts, it is usually possible to recognize which indicators are highly correlated, and to select one that can represent a larger set.

A set of selected indicators could be compiled into an Index of Individual Development as a supporting tool for a New Social Contract. These could be measured as the percent of the population with the indicator, and/or the quality of the service or government performance in assuring that dimension for everyone in the country. Some sample indicators for each of the dimensions of individual development are shown in Table 3.

Table 3 Sample indicators by dimensions of development

Physical growth/health
Healthcare provision for the birth of a child
Accessing primary health care (cost and access to primary health care facilities)
Adequate child nutrition (government programs and institutions that ensure adequate and nutritious food for the growth of every child)
Access to clean water and basic sanitation
Legal and institutional frameworks to assist the elderly
Material welfare indicators
Number of homeless people or families living in inadequate and unsanitary conditions
Access to on or off-grid electricity (including the time, steps and cost)
Meeting basic energy needs for cooking (access, cost, and health and environmental impacts)
Installing solar energy at home (availability, regulations, time and costs for domestic solar water heater or photovoltaic panels)
Security and safety
Access to insurance or other protection against disaster or theft
Availability of protection and support to women victims of domestic violence
Managing toxic chemicals (government environmental regulation in the public interest)
Legal protection (workers compensation and consumer protection legislation, and efficiency of local courts)
Causes of premature death in adults, including (a) acts of aggression, (b) work-related deaths, (c) traffic accident mortality, (d) deaths from natural hazards (drought, famine, earthquakes, storms, floods) etc.
Education
Access to school for every child to learn to read, write and calculate
Enrolling in school (steps, time and cost for enrolling a 7 year old girl into primary public education)
Percentage of (a) Urban boys (b) Urban girls (c) Rural boys (d) Rural girls that successfully complete primary education with basic literacy skills
Percentage of functional literacy in the country by gender, age group and class, caste or ethnic group
Work
Proportion of the population that works in: (a) the cash economy, (b) the subsistence economy, (c) the informal sector, (d) the illegal sector, or (e) non-waged work like housewife or raising a family
Legislation related to getting a job (worker protection measures and gender parity in employment regulations)
Government assistance to school leavers to find a job, craft or subsistence activity
Financial system
Buying a residence (regulatory environment for housing finance)
Opening a savings account (ease of financial transactions)
Paying taxes (measuring the payments, time and total tax rate for a typical individual)
Experience of corruption (payments to receive government services, bribes to avoid taxes or fines, receiving only partial entitlements)
Government provision of social security during unemployment, illness, or if handicapped
Government protection of the value of earnings, capital, savings and pension against inflation

(continued)

Table 3 (continued)

Justice and fairness
Registering a newborn with country authorities (time, steps and cost)
Having an official household address (a key barrier to identification in many developing countries)
Getting married (capturing the bureaucratic red tape involved in complying with licensing regulations around marriage)
Getting a passport (regulations, requirements, time and cost to obtain the document necessary for international travel)
Human rights and freedoms
Protection of freedom of speech and right to receive and impart information
Freedom of thought, conscience and religion, and to change religion or belief
Right to privacy of person, family, home, correspondence
Right to own property
Free movement and choice of place of residence
Right to take part in government, to vote, to participate in political life
Place in the community
Recognition of personal status and dignity
Presence of social and family networks
Community trust, reciprocity and resilience
Registering to vote (possibility of participating in public elections)
Getting a cell phone (mobile) and regular connection (cost and procedures)
Obtaining Internet access (cost, regulations and possible censorship)
Accessing public transport (availability, time and cost)
Cultural and spiritual identity
Legal protection of cultural identity and heritage
Presence of a value system, beliefs, ethics and morals
Having a vision and purpose in life, hope for a better world
Opportunities for participation in culture and the arts
Access to beauty, to nature
Overall evaluative well-being or life satisfaction

Values-Based Indicators of Individual Motivation

In addition to indicators of dimensions of human well-being, it can be useful to develop indicators of individual motivation to improve individual well-being and to contribute to the well-being and advancement of the group or community. Indicators of behaviors or attitudes can help individuals become more aware of what their values are, and where there might be inconsistencies between different values, or between their values and their behavior or lifestyle. They get closer to the root causes of an individual's unsustainable lifestyle, and can have an emotional impact with the power to motivate change. The aim is to increase self-awareness of one's positive sustainability values and to encourage their development, while signaling areas where improvement is needed.

Values-based indicators can be self-assessed through questionnaires, or measured through a variety of social science assessment techniques, such as semistructured interviews, participant observation, focus groups, etc. The most useful indicators can measure individual behaviors and attitudes associated with these values. For example, a value statement could be assessed simply agree/do not agree, on a scale (not important-->very important) or by selection from a choice of statements expressing a range of feelings about an item: (i.e.: I hug trees for spiritual strength. Trees inspire and refresh my spirit. Trees produce the oxygen I breathe. Trees give me wood and paper. Trees are nice, but things are cleaner without them. When you have seen one tree, you have seen them all.)

The following values have been identified as relevant to motivations towards sustainable development, adapted for individual self-assessment (Podger et al. 2013; Dahl 2012b; http://www.esdinds.eu):

- Respect for the environment
- Empowerment
- Appreciating unity in diversity
- Trustworthiness/integrity
- Justice/solidarity
- Moderation and detachment from material things.

Respect for the Environment

These indicators assess whether the individual sees himself/herself as separate from the environment, or as part of it, dependent on it, and intimately linked to the natural processes of the biosphere and to the organisms with which we share it. Many indigenous peoples saw no separation between them and their surroundings. Many religions teach stewardship for God's creation. Science has demonstrated our environmental dependence and vulnerability in many ways. Regardless of its origin, the result is a desire not to cause environmental damage.

Examples of Indicators

- I value the natural world as a source of personal fulfillment
- I purchase environmentally sustainable products even if cheaper alternatives exist
- I try to make my recreation, social activities and celebrations environmentally friendly

Empowerment

Empowerment reflects the ability to act and to make a difference, an awareness that many drops can make an ocean. Its absence is a frequent cause of lack of motivation. In a community or educational situation, it can be the result of encouragement and accompaniment, and other positive reinforcement.

Examples of Indicators

- I am encouraged to grow personally and reach my potential
- I feel that others respect me, as I respect them
- I can take risks, make mistakes and learn from my errors
- I do not have to compromise my personal beliefs or values
- I feel that I am able to effect change

Appreciating Unity in Diversity

The greater the feeling of unity between individuals or within a community, the greater their power to work together and to overcome differences. This reinforces teamwork and encourages innovation, since it includes the recognition that there can be many different solutions appropriate to different situations, and that not everyone has to do the same thing to achieve a shared goal. It inspires confidence in one's own abilities. It also overcomes prejudice and facilitates the appreciation of others.

Examples of Indicators

- I listen to and respect other people's points of view
- I try to ensure that everyone is included
- I replace a negative feeling towards someone by a stronger positive feeling
- My community is richer because of its diversity

Trustworthiness and Integrity

Trust is essential to any positive group interaction. Someone who has integrity will rapidly be integrated into a group. Trustworthiness is the result of consistency between words and actions, and an inner as well as outer honesty. Cultivating these values facilitates group interactions, brings respect, and motivates consistency in sustainable behaviors.

Examples of Indicators

- I am honest and meet my obligations even when there is no chance of being caught
- I follow through with my commitments
- I try to practice what I preach; my actions are consistent with my words
- I can be trusted with other peoples' money

Justice and Solidarity

Justice has already been highlighted as the essential foundation of a new social contract. It is also an important individual value, as it allows a person to see with his or her own eyes and not through the views of others. It has the power to tip the balance between self-interest and the common interest or the interests of others so necessary to achieve sustainable behaviors.

Examples of Indicators

- I identify what is right for myself and do not rely on the opinions of others
- I try to help those less well-off than myself
- I take into account the needs of future generations
- I give voluntarily to support social causes
- I pay all my taxes

Moderation, Detachment from Material Things

For those tempted by the consumer society, these values can be a good guide to responsible living. Developing the capacity to distinguish needs from wants is an important step towards sustainable consumption and resistance to commercial manipulation. Again, these values are at the root of many spiritual traditions, and contribute to physical and mental health.

Examples of Indicators

- I only buy what I really need
- I place no importance on status symbols
- I prefer to invest in social relationships rather than material goods
- I believe that wealth can be a barrier to spiritual development

Conclusions

When sustainable development is considered in the wider context of human purpose and well-being presented here, it takes on a whole new meaning, in which its economic, social and environmental dimensions are fully integrated. At the same time, rather than seeming utopian and unattainable. It is precisely this emphasis on the social, cultural, ethical and spiritual aspects of well-being that can motivate changes in human behavior and drive a bottom-up

(continued)

transformation in human society across all its many and diverse components. This is essential to eliminating poverty through empowerment and increased capacity rather than the unsustainable charity of development assistance. The focus on the individual makes sustainable development immediately relevant. While global environmental problems and failures in economic and political systems may seem remote from individual concerns and possibilities of action, everyone can start to act to bring improvements in their relations with others within their local community and work-place, and to experience the self-reinforcing effect of visible results in improved well-being.

Creating new measures of development at the individual level will help to change the focus from creating wealth to creating well-being in a spirit of justice and equity. In addition, values-based indicators can make people conscious of their real desires and motivations, and build an emotional commitment to change. By getting the signaling right, we can measure implementation of the social contract for a just and sustainable society that enables every human being to fulfill his or her potential in life both in cultivating individual qualities, personality and capacities and in contributing to the advancement of civilization.

References

Baha'í International Community (2010) Rethinking prosperity: forging alternatives to a culture of consumerism, Bahá'í International Community's contribution to the 18th session of the United Nations Commission on Sustainable Development, http://bic.org/statements-and-reports/bic-statements/10-0503.htm

BIC (1995) The prosperity of humankind. Office of Public Information, Haifa

BIC (1998) Valuing spirituality in development: initial considerations regarding the creation of spiritually based indicators for development, a concept paper for the world faiths and development dialogue, Lambeth Palace, London

Dahl AL (1996) The eco principle: ecology and economics in symbiosis. Zed Books/George Ronald, Oxford

Dahl AL (2012a) Achievements and gaps in indicators for sustainability. Ecol Indic 17:14–19. doi:10.1016/j.ecolind.2011.04.032

Dahl AL (2012b) Ethical sustainability footprint for individual motivation. Planet under pressure 2012 conference, London. http://iefworld.org/ddahl12d

Helliwell J, Layard R, Sachs J (2012) World happiness report. Earth Institute, Columbia University, New York, http://www.earth.columbia.edu/articles/view/2960

Maslow AH (1943) A theory of human motivation. Psychol Rev 50:370–396, http://psychclassics.yorku.ca/Maslow/motivation.htm

Podger D, Velasco I, Luna CA, Burford G, Harder MK (2013) Can values be measured? Significant contributions from a small civil society organization through action research. Action Res J 11(1):8–30

Rawls J (1999) A theory of justice, Rev edn. Harvard University Press, Cambridge

Rockström J, Steffen W, Noone K, Persson A, Chapin FS, Lambin E III, Lenton TM, Scheffer M, Folke C, Schellnhuber H, Nykvist B, De Wit CA, Hughes T, Van der Leeuw S, Rodhe H, Sörlin S, Snyder PK, Costanza R, Svedin U, Falkenmark M, Karlberg L, Corell RW, Fabry VJ, Hansen J,

Walker B, Liverman D, Richardson K, Crutzen P, Foley J (2009) Planetary boundaries: exploring the safe operating space for humanity. Ecol Soc 14(2):32, http://www.ecologyandsociety.org/vol14/iss2/art32/
Stiglitz JE, Sen A, Fitoussi JP (2009) Report by the commission on the measurement of economic performance and social progress. http://www.stiglitz-sen-fitoussi.fr
UN High-level Panel (2013) A new global partnership: eradicate poverty and transform economies through sustainable development, the report of the high-level panel of eminent persons on the post-2015 development agenda. United Nations, New York
UNDP (2014) 2014 human development report: sustaining human progress: reducing vulnerabilities and building resilience. United Nations Development Programme, New York http://hdr.undp.org/en/
United Nations (1948) The universal declaration of human rights. http://www.un.org/en/documents/udhr/
United Nations (2012) The future we want, annex to resolution 1 in report of the United Nations Conference on Sustainable Development (A/CONF.216/16), Rio de Janeiro, Brazil
United Nations (2013) The millennium development goals report 2013. United Nations, New York
Ura K, Alkire S, Zangmo T, Wangdi K (2012a) A short guide to Gross National Happiness Index, Thimphu, Centre for Bhutan Studies, Bhutan. Available from Bhutan Gross National Happiness Commission. http://www.gnhc.gov.bt/
Ura K, Alkire S, Zangmo T, Wangdi K (2012b) An extensive analysis of GNH index, Thimphu, Centre for Bhutan Studies, Bhutan. Available from Bhutan Gross National Happiness Commission. http://www.gnhc.gov.bt/
World Commission on Environment and Development (1987) Our common future. Oxford University Press, Oxford

Insights for a Better Future in an Unfair World: Combining Social Justice with Sustainability

François Mancebo

Abstract One of the more challenging aspects of sustainability policies is to address social justice. Often, so-called sustainability initiatives turn out to be completely out of touch with the needs and expectations of the populations concerned, and contribute to increase social injustice. This is particularly true in urban areas, which are the ambit of this chapter. How to cope with this problem? Promoting people's place-based appropriation of sustainability policies looks like an interesting lead to follow. The challenge here is to address the social process of decision-making. Ultimately, the challenge is design a new social contract: matter in which a comprehensive understanding of the coordination mechanisms between the local, national, regional and international scale is crucial.

Keywords Transition to sustainability • Social process of decision-making • Urban planning • Multilevel governance • Imported sustainability

Achieving a livable and sustainable future in a changing world is a crucial challenge that our societies are facing. On this point, everybody agrees. Though, when it comes to determining how to do this practically, or simply what sustainability really is about, and, there is much less consensus. Addressing the antagonisms between social justice and sustainability is a way amongst many to address transition to sustainability. This chapter focuses on the sharp processes of spatial differentiation and the many-fold conflicts between urban sustainability and social justice.

The reason why it is so difficult to answer the basic question of what sustainability is lays into the fact that sustainable development is not only about science. It also is about ideas and values (Leiserowitz et al. 2006): various interpretations—frequently

---

François Mancebo is Full Professor of Planning and Sustainability at Rheims University. He is the head of the IRCS (International Research Center on Sustainability) and the director of the IATEUR (Institute of Regional Development, Environment and Urban Planning of Rheims). Email. francois.mancebo@univ-reims.fr

F. Mancebo (✉)
IATEUR IRCS, University of Reims Champagne-Ardenne, Campus Croix-Rouge, 57 bis avenue P. Taittinger, 51100 Reims Cedex, France
e-mail: francois.mancebo@univ-reims.fr

F. Mancebo, I. Sachs (eds.), *Transitions to Sustainability*,
DOI 10.1007/978-94-017-9532-6_9

divergent from one another—naturally thrive, since values may differ a lot between cultures and over time (Christen and Schmidt 2012). Starting with the release of the World Commission on Environment and Development report (WCED 1987), sustainable development began being widely discussed throughout the 1990s among international organizations. Rio's Agenda 21 has seen national and even regional and local governments enter the debate after 1992 (Eliott 2006). Progressively, sustainable development took on a multiplicity of sometimes-contradictory meanings (Robinson 2003). The situation is now so complex that it needs a mapping (Hopwood et al. 2005). However, the central idea is simple enough: Recognizing the finite nature of earth's biophysical resources, it promotes a type of development that meets our current needs without compromising those of the generations to come. When the United Nations assigned the redaction of a report to the World Commission on Environment and Development, its mission statement mentioned explicitly that its objectives were how to reduce inequality and poverty without damaging the environment granted to the future generations (WCED 1987). Such a statement entails serious difficulties: It is not so easy to balance the needs of our societies today (environmental justice, living conditions) with the needs of the future generations (preservation of the resources and protection of the planet).

This difficulty cuts across another one of the same nature: the theoretical divide between "weak" and "strong" sustainability (Haughton and Hunter 1994). Proponents of "weak" sustainability consider that manufactured capital can replace completely natural capital, as technology answers the environmental consequences of the production of goods and services: "the world can, in effect, get along without natural resources, so exhaustion is just an event, not a catastrophe" (Solow 1974). Proponents of "strong" sustainability consider that manufactured capital cannot replace perfectly natural capital, especially some global processes vital to the human existence such as the climate or the ozone layer (Daly 1998; Roseland 1998). In this perspective, it is crucial to limit the quantities of material and energy extracted from the biosphere and to reduce drastically the emission of pollutants (Von Weizsäcker et al. 1997; Lenton et al. 2008). Officially, of course, sustainable development is an integrative notion that should harmoniously unify these two types of sustainability (Giddings et al. 2002). The Brundtland report points out that the satisfaction of human aspirations should "*not endanger the natural systems that support life on Earth: the atmosphere, the waters, the soils, and the living beings*"…"*It is part of our moral obligation to other living beings and future generations*" (WCED 1987). But, simultaneously it promotes a more rapid economic growth in order to overcome poverty, in reference to the "trickle-down theory" which affirm that the economic grows finally benefit to everybody and as such reduce poverty (Dollar and Kraay 2000). Such an ambiguous position is of no help to clarify the link between the two types of sustainability. It creates confusion on what is the substance of sustainable development. Thus, it generates the gap between "weak" and "strong" sustainability, which finally is a gap between those who give poverty reduction and social justice today priority over the needs of the generation to come (which calls for "weak" sustainability), and those who think the opposite (which calls for "strong" sustainability).

How Sustainability May Fosters Injustice: A Focus on Urban Areas

The challenge of combing social justice with sustainability policies is particularly crucial in urban areas. By 2050, 70 % of the world's population will live in cities (UN-HABITAT 2008). Local authorities, when trying to make their city sustainable, have in common the objective of a better use of what is already there (Theys and Emelianoff 2001). Sustainable cities could thus be nicknamed "recyclable cities" in the sense that they are supposed to constantly recycle their urban fabric and their urban functions without going through phases of obsolescence with brownfield land and degraded neighborhoods, and without squandering soils (Swart et al. 2003). For example, new "ecological gardens" appeared at the end of the 90s, on the wastelands of former industrial sites of inner Paris and at the same time old industrial buildings of these sites were rehabilitated in eco-friendly construction (apartments or offices) (Duréault 2013). But even if recycling partly what is already there, these cities will have to provide water and energy to their inhabitants while reducing pollution and using sustainable resources. A challenge that requires radically new procedures and technical tools to manage traffic congestion, water and electricity networks, "intelligent buildings," while preserving the existing urban and social fabric. Therefore, urban areas are relevant research objects when trying to examine how to combine social justice with sustainability policies.

As shown by Elizabeth Burton in a large sample of UK towns, technical solutions may join with legal requirements in increasing social injustice (Burton 2001). Consider the particular case of sustainable housing: As far as sustainable housing is concerned, the reason why sustainable cities and ecological neighborhoods are mostly inhabited by wealthy people is simple: In the beginning, these categories were targeted because they could afford the higher construction costs, and because they were decisive in the formation of new trends. Such a choice was supposed to make easier the democratization of the access to these type of living, as larger demand would made possible lowering construction costs due to economies of scale. The Swedish cases of *Hammarby Sjöstad* (Stockholm) or *Västra Hamnen* (Malmö) illustrate this approach (Olander et al. 2007). However, this democratization did not happen. Construction costs inflated steadily, as developers, constrained by drastic environmental specifications, played the "style and class" card to increase their capital gains. Anyway, as higher as it can be, there are a limited number of ecological dwellings, and their attractiveness is strong. So, the law of supply and demand increases the rent rate and the sell rate, regardless what the construction cost is. This new upward pressure on prices brought by sustainable housing usually proves catastrophic (François et al. 2011). The name of "sustainable" neighborhood is inappropriate when a neighborhood becomes socially inaccessible. This leads some authors to denounce the veil thrown over profoundly unfair environmental dynamics that involve the departure of socially vulnerable people out of these places to outlying areas (Smith 2002).

There is another issue concerning social justice here: wouldn't public money have been spent more efficiently if invested to reduce actual environmental disparities between areas, bettering places where the environmental conditions are already pretty bad? In the metropolitan region of Paris, for example, 50 % of the places with degraded environment (pollution, nuisances) are also socially deprived; symmetrically, nearly 50 % of those with good environmental conditions are wealthy areas (Bigot 2009). If we try to consider what is the main factor for such a distribution, the attractiveness of the *communes* with a nice environment appears less decisive than the avoidance of the nuisances of those with a poor one. What is interesting here, is that the residential choice is motivated by the rejection of environmental degradation rather than the attractiveness of environmental amenities (nature, silence, air and water quality, etc.) (Gueymard and Faburel 2008). Thus urban sustainability policies should focus on an inclusive approach, rather than to keep on creating "attractive" green housing spots haphazardly. More generally, sustainability policies should be conceived and implemented in areas large enough to take into account not the whole urban fabric.

The Issue of Imported Sustainability

When a place looks sustainable by giving to other places the burden of its transition to sustainability—exporting pollution and undesired products (waste and nuisances) or polluting activities, siphoning their resources and energy—this place is not really sustainable. It benefits from what David Pearce calls imported sustainability (Pearce et al. 1989), that is to say in the case of a urban area when a city transfers the cost of its sustainability onto adjacent or distant regions. Sometimes imported sustainability is an unintentional phenomenon, for example, fires in boreal peat lands may be sources of atmospheric mercury, transported and deposited far away (Turetsky et al. 2006).

Imported sustainability is a major bias against the implementation of sustainability policies. An effective sustainability policy should be conceived on an area large enough to internalize the imported sustainability bias, while taking into account all the relations between the human beings and the environment where they live (Elliot 2006). In the case of urban policies the only solution is defining them on extensive spatial scales, which include suburban, periurban and dependent rural, or natural areas (Donzelot 2004). It is all but evident, since limits will differ according to which aspect of sustainability we focus on: The functional area and the employment area of a major industrial center do not coincide, nor do they with the geographical area affected by the pollution (physical, chemical, air and water) and nuisance due to this industrial center. Thus, to avoid imported sustainability, urban sustainability policies should be conceived and implemented at three complementary scales simultaneously:

- First is the scale of the neighborhood. A place based level. At this level the physical impact of the urban projects, even if they are conceived at the agglomeration level, is maximal.
- Second is the scale of agglomeration in urban planning. Which represents finally a cluster of adjacent neighborhoods working together. It gives a good insight of the urban policies, on the one side, and of the urban lifestyles, on the other side. This level plays a strategic role in sustainable urbanization. At this scale the coordination between multiple actors producing policies is crucial.
- Finally, there is the scale of the hinterland, which reflects the agglomeration environmental footprint. It is defined to include most of the fluxes of the urban metabolism (Billen et al. 2012). This level can be called "regional." It is crucial to describe imported sustainability, since it is supposed to encompass a significant part of the environmental footprint.

Determining concretely these three scales is tricky. Urban areas are covered with overlapping partitions: Each administration, each economic actor, each local community produces its own zoning and its own limits. So-called sustainability policies can have terrible effects when they do not take into account scales linkage.

What Is a "Good" Environment? A Place-Based Perspective

The context is important when cascading through spatial scales. Every person and community lives multiple affiliations, based on various territorial scales. Thus, sustainability policies must address the existing social a cultural fabric, legislation and planning traditions, communities, local assets and resources (Costanza et al. 2001). Thus to combine social justice with sustainability, it is fundamental to understand the linking between the societies and the ecosystem where they live in at different spatial scales (Carpenter et al. 2009). One of its expressions is spatial heterogeneity. For example, patterns of land use/land cover strongly influence hydrologic flow paths and delivery of nutrients to surface waters; patterns of agricultural and natural/semi-natural habitats affect the diversity and abundance of natural enemies that prey upon agricultural pests, etc. (Strayer et al. 2003; Werling and Gratton 2008). Humans often re-scale spatial patterns, increasing heterogeneity at large scales while reducing heterogeneity at small scales. For example, in agricultural areas, humans often impose coarse spatial patterns with sharp boundaries and greater contrast among land covers while homogenizing fine-scale variation in soil properties. The sharp boundaries, high contrast, and altered functional connectivity resulting from human activity may change the quantity, quality and variability of landscapes (Turner et al. 2008). In addition, this spatial heterogeneity reflects heterogeneity among people, cultures and institutions that affects sustainability and social justice (Turner 2010). When considering this, societies appear as complex adaptive systems, composed of individual agents who have their own priorities, and who value the macroscopic features of their societies differently. Resolving those competing perspectives is at the core of transition to sustainability. Complex adaptive systems, integrate change from individuals to whole systems, across scales. The recurring question of which coordination mechanisms are needed at the local, regional, national or international scale is central here, especially to meet the needs of policy-makers for decision-making (Carpenter 2010). In such systems, macroscopic patterns emerge, to large extent, from interactions at much lower scales of organization – individual agents, short time scales, and small spatial scales- and feed back to influence the dynamics at those microscopic scales, as assessed by Simon Levin (Levin 1992, 2010).

To understand and cope with the outcomes of such complex human environment systems, the contrasts across locations are particularly important (Daily et al. 2009). Thus a place-based approach is fundamental for a sound transition to sustainability. It is therefore essential, when trying to combine social justice with sustainability, to determine locally what is a good environment for the communities involved: One in which the improvement of environmental conditions *stricto* sensu (water quality, air, biodiversity, prudent use of resources, land and energy, etc.) will lead to improved living conditions; one in which technical devices and ecological processes—included in areas large enough to take into account imported sustainability—will lead to new lifestyles.

There is a gap, for example, between real environmental nuisance and its perception through the notion of quality of life (Moser and Weiss 2003). It should be noted, for example, that French *espaces verts* (green areas), do not necessarily bring people together. They also isolate people because their separate their homes. This aspect is in line with the Parisian history: the introduction of greenery by Haussmann was an attempt to control the use of public space by a technical approach based on hygienism (Luginbuhl 1992). Its main function was to bring more sunlight to the city and better the air circulation. The city life was marked by socio-spatial differentiation, virtually segregative, embodied in a type of revegetation reduced to *espaces verts*. Its role is finally to separate, to distinguish and to hide (Moret 2004). The very term *espace vert* (green area) reveals its real nature: "… by losing its name, the old urban garden or urban park is deprived of its positive attributes… the *espace vert* is no longer a place but rather an indistinct area whose boundaries are decided in the abstract world of the master-plans…" (Le Dantec and Le Dantec 1987). The current of Paris regional master plan proposes—as an important mean to foster sustainability—a quantitative objective of 10 m^2 of public green area per inhabitant at the communal level. As though it were sufficient to display "green" to become suddenly sustainable.

Two Gordian Knots: Intergenerational/Spatial Equity, Weak/Strong Sustainability

In urban planning, one among the many challenges of sustainability is reestablishing the inclusiveness of the urban and social fabric—which is a complex task—instead of popping-up buildings or housing estates without paying attention to the surroundings—which is so easy—The shape and outline of the cities, their *vela*, compose their urban form and determine their identities as well. To foster a good quality of life, there is need for contrasts, to meet and to adapt to the different individual aspiration among the inhabitants. Urban reconversion is crucial here: For example, industrial wastelands in the inner suburbs may be converted into offices or apartments, as part of eco-neighborhoods. Such sustainable actions are supposed to integrate urban *habitus* into the new projects.

But, more than often, things turn out very differently: "Exemplary" buildings and devices (Willbanks 2003)—all technical solutions—are often favored to the detriment of more holistic approaches, such as active land management and transformation of the urban fabric (differential densification, restructuring urban cores, etc.). To promote "green" buildings, elected officials accept to pay extra charges up to 20 % of the original costs to obtain a Low-Energy label. They are less interested in the urban design, which is more important to create a real sustainable city but, of course, harder to implement and less profitable as an electoral issue. Besides, mayors, representatives and more generally elected officials adore showcasing constructions and they love them "brand new." They are so much more visible. Thus, too often, developers deliver turnkey new energy efficient construction and passive buildings in new neighborhoods improperly called "environmentally friendly" (Bierens de Haan and Dawson 2006). In many cases, vegetation, green technologies and exterior wood facing, camouflage very classical housing estates totally disconnected from their surroundings. Naturally, the regeneration of the existing urban and social fabric is not addressed here. There is no way to foster communities in such a context. The identity of place is usually extraordinarily weak for the people living there (Proshansky et al. 1983).

That kind of mechanism is the main reason why sustainability policies have finally few public backing, and are perceived unfair and technocratic. It explains the failure of numerous so-called sustainability actions to meet their target whether social or environmental: the people concerned do not take ownership of them. We should never forget that eventually, it is the current populations and societies that decide what is a "good environment," not the future generations who are not already here to push their ideas about what is a "good environment?" This bias speaks to us of the dilemma between preserving the environment for the generations to come—what we can call intergenerational equity—and prioritizing actual issues, such as quality of life or social justice—what we can call geographical equity—. There is a general equity principle, which we could also call fairness, at the heart of sustainable development (Cairns 2001). But in fact, there are many equities (Gibson et al. 2005). Usually, academic authors differentiate between intergenerational equity, geographical equity, procedural equity and, finally, interspecific equity (Haughton 1999). But in fact, the confrontation between intergenerational equity and geographical equity is what structure most strongly sustainability policies—as seen previously with the opposition between "weak" and "strong" sustainability—especially by urging on a better articulation between short-term (geographical equity, including social justice) and long-term (intergenerational equity). To make reference to Amartya Sen (Sen 2009), if there are obligations toward future generations, there are also obligations toward the actual generation. To combine sustainability issues and social justice, it is thus necessary first to make sustainability policies acceptable to the current populations, and naturally these populations are prone to favoring their interests—here and now—to issues placed in a distant future.

Participatory Joint-Construction of Sustainability Policies

Since, the effectiveness of sustainable policies lies largely in their acceptability—a highly subjective and rarely disinterested matter (Fischhoff et al. 1981)—and in their collective ownership, decision-making processes should be fundamentally a matter of collective decision. Beyond their procedural and prescriptive appearances, these decisions result from the confrontation—or the synergy—of choices made by a myriad of actors, each acting for its own concern and its own world vision. The more adequate framework to address such a situation is participatory joint-construction of these policies (Andrews 2002): A boundary work, since it uses knowledge to inform negotiation among relevant actors in a politicized context; which corresponds to "political bargaining," according to the definition of William Clark (Clark et al. 2011), where actors with their own interests interact with heterogeneous knowledge producers. The point is co-producing collective decision through the interaction between society and science (Jasanoff and Wynne 1998), in an attempt to legitimize sustainability policymaking. This should include non-market organizations, local communities and individuals able to form self-determined user associations, in the continuity of Elinor Ostrom's work that showed that user communities with neighborhood governance could manage common resources more efficiently than the market or institutional structures (Ostrom 1998). There are three main obstacles:

- First, it is difficult to encompass all the actors (regional and local authorities, non-market institutions, NGOs, private companies, local store keepers, unions and chambers, landowners, etc.), even more to visualize the whole of their interactions.
- Second, how to take into account in the process the micro-decisions made by individuals and households, which have an indirect but strong influence on collective decisions. They are shaped by the moment and the economic status of the persons: Depending on whether—at moment t—they feel (or are really) poor or not, they will not make the same choice if they are placed in the alternative of eating properly or going to the theater, thermally insulating their house or paying their bills. Ostentatious choices also play a big role, since they determine their position on the "social totem" (Frank 1999). Thus, to which point having a house of 1,500 m^2 gives you more happiness than one of 1,000 m^2? Not much more (Winkelmann 2012), but you need to "keep up with the Joneses" to conform with the social codes (Drakopoulos 2013), and because the demand is there, the size

of the houses keeps rising inexorably on, accelerating the urban sprawl while denying more and more people the ability to house themselves properly.

- Third, all the actors have to consider the other members as legitimate partners, and the process of co-construction itself as satisfying the criteria of saliency, credibility, and legitimacy, which is all but evident to achieve (Mollinga 2008). Indeed, decision makers can be tempted to use the workshops to support decisions they have already made, or avoid responsibility by repackaging them as technical issues to be resolved by experts they control (Weingart 1983). Besides, how to convince each member of the panel that the content of the workshop is not biased in support of another member's agenda (Van Noordwijk et al. 2001).

To avoid these obstacles, it is necessary to pay great attention to two points:

- The process itself has to be flexible enough to be meaningful for actors coming from different "social worlds" (Jasanoff 1987; Turnhout 2009). As such, they necessarily exhibit a certain degree of vagueness and ambiguity, while maintaining consistency.
- There should be a focus on identifying the linkage between collective and individual decisions. By collective decisions, I mean decisions made by organizations or institutions (officials or not): urban form, transport policy and so on. By individual decisions, I mean the decisions taken by individuals or households autonomously, but that will aggregate to affect the collective decisions, while being influenced by them.

Conclusions

Combining the increase of everyone's wellbeing and social justice with sustainability is one of the greatest challenges today. To do so, sustainability policies, should focus more on the social process of decision-making. It means considering that the environment, far from being pure transcendence, is embedded in the societies. Quiet and nice unpolluted living environments have become *emblèmes* in the sense of Pierre Bourdieu and, as such, highly attractive—and expensive—areas. The human being builds a representation of the ecosystems he lives in and calls it "environment," out of the usages he makes of their resources: Takings (usage of air, water, minerals), inputs (pollution), alterations (housing, transport). Thus, the environment represents a more or less noisy neighborhood to which we have to adapt. A polluted environment can be a place where life is good. Conversely, an environment with clean air and clean water can be quite intolerable as evidenced by windswept segregated social-housing complexes settled in the middle of nowhere, where the quality of life is low (Mancebo 2010). It is impossible to determine whether a place is sustainable or not only by considering the factual date of environmental indicators. Instead, sustainability is an inclusive notion, which integrates social, cultural and economic aspects of the concerned societies.

(continued)

Promoting collective appropriation of sustainability policies implies that those who will be affected by them are involved in the process of decision-making, right from the beginning. When they are disconnected from the inhabitants and local communities needs, desires and definition of what a "good environment" is, these policies fail to meet their objectives. In such a context, it is crucial to design a new social contract that include stakeholders, neighborhood communities and groups of individuals among the major actors of sustainable development. This social contract is therefore and first of all a political process, which has to meet the following issues: What type of society do we want to live in? Which compromises between the goals and interests of the different groups? What linkage between one decision level and another?

The Millennium declaration proclaimed the "collective responsibility to uphold the principles of human dignity, equality and equity at the global level" (Stokke 2009). Prior to that, when the United Nations assigned the redaction of a report to the World Commission on Environment and Development in 1983, which is the source of sustainable development, its mission statement mentioned explicitly that its objectives were: "How to reduce inequality and poverty without damaging the environment granted to the future generations" (WCED 1987). It is time to go beyond the mantra, and try to do it concretely.

References

Andrews CJ (2002) Humble analysis: the practice of joint fact-finding. Praeger, Westport

Bierens de Haan C, Dawson J (2006) Entre écovillages et projets d'architectes: les écoquartiers. L'Urbanisme 348:41–44

Bigot R (2009) Regards sur les quartiers sensibles et les discriminations en France, Agence nationale pour la cohésion sociale et l'égalité des chances, Collections des rapports, n° 271, CREDOC, Paris

Billen G, Barles S, Chatzimpiros P, Garnier J (2012) Grain, meat and vegetables to feed Paris: where did and do they come from? Localizing Paris food supply areas from the eighteenth to the twenty-first century. Reg Environ Chang 12:325–335, Springer

Burton E (2001) The compact City and social justice. Housing, environment and sustainability, Housing Studies Association Spring Conference, University of York, York UK

Cairns J (2001) Equity, fairness, and the development of a sustainability ethos. Ethics Sci Environ Polit 1:1–7

Carpenter S (2010) Measuring and monitoring progress toward sustainability. In: Levin SA, Clark WC (eds) Toward a Science of Sustainability. CID Working Paper, n° 196, pp 135–139, Harvard

Carpenter SR, Mooney HA, Agard J, Capistrano D, De Fries R, Diaz D, Dietz T, Duriappah A, Oteng-Yeboah A, Pereira HM, Perrings C, Reid WV (2009) Science for managing ecosystem services: beyond the millennium ecosystem assessment. Proc Natl Acad Sci 106:1305–1312

Christen M, Schmidt S (2012) A formal framework for conceptions of sustainability–a theoretical contribution to the discourse in sustainable development. Sustain Dev 20(6):400–410

Clark WC, Tomich TP, Van Noordwijk M, Guston D, Catacutan D, Dickson N, McNie E (2011) Boundary work for sustainable development: natural resource management at the Consultative Group on International Agricultural Research (CGIAR). Proc Natl Acad Sci. doi:10.1073/pnas.0900231108
Costanza R, Low BS, Ostrom E, Wilson J (2001) Institutions, ecosystems, and sustainability, Ecological economics series, Lewis Publishers, Albany GA
Daily GC, Polasky S, Goldstein J, Kareiva PM, Goldstein LP, Ricketts TH, Salzman J, Shallenberger R (2009) Ecosystem services in decision making: time to deliver. Front Ecol Environ 7:21–28
Daly H (1998) Reconciling internal and external policies for sustainable development. In: Dragun AK, Jacobson KM (eds) Sustainability and global economic policy. Elgar, Cheltenham
Dollar D, Kraay A (2000) Growth is good for the poor. World Bank, Washington, DC
Donzelot J (2004) La Ville à Trois Vitesses: Relégation, Périurbanisation, Gentrification. Revue Esprit 3–4:14–39
Drakopoulos SA (2013) Hierarchical needs, income comparisons, and happiness levels. Posit Psychol Perspec Qual Life Soc Indic Res Ser 51:17–32
Duréault J (2013) Architecture contemporaine et nature en ville, Engineering Thesis, Agrocampus Ouest, Angers
Elliot J (2006) An introduction to sustainable development. Routledge, London
Fischhoff B, Lichtenstein S, Slovic P, Derby S, Keeney R (1981) Acceptable risk. Cambridge University Press, Cambridge
François J-C, Ribardière A, Fleury A, Mathian H, Pavard A, Saint Julien T (2011) Les disparités de revenus des ménages franciliens – Analyse de l'évolution entre 1999 et 2007, Collection Présent pour avenir, DREAL d'Ile-de-France
Frank R (1999) Luxury fever: money and happiness in an era of excess. Princeton University Press paperback. The Free Press, Washington DC
Gibson R, Hassan S, Holtz S, Tansey J, Whitelaw G (2005) Sustainability assessment—criteria and processes. Earthscan, Oxford
Giddings B, Hopwood B, O'Brien G (2002) Environment, economy and society: fitting them together into sustainable development. Sustain Dev 10:187–196
Gueymard S, Faburel G (2008) Vécu environnemental et qualité de vie en Région Ile-de-France: une approche exploratoire des inégalités environnementales. PUCA, Paris
Haughton G (1999) Environmental justice and the sustainable city. J Plan Educ Res 18:233–243
Haughton G, Hunter C (1994) Sustainable cities. Kingsley, London
Hopwood B, Mellor M, O'Brien G (2005) Sustainable development: mapping different approaches. Sustain Dev 13:38–52
Jasanoff S (1987) Contested boundaries in policy-relevant science. Soc Stud Sci 17:195–230
Jasanoff S, Wynne B (1998) Science and decision-making. In: Rayner S, Malone EL (eds) Human choice and climate change: the societal framework. Battelle Press, Colombus OH, pp 1–87
Le Dantec D, Le Dantec J-P (1987) Le roman des jardins de France, éditions Terre de France, Plon
Leiserowitz A, Kates R, Parris T (2006) Sustainability values, attitudes, and behaviors: a review of multinational and global trends. Annu Rev Environ Resour 31:413–444
Lenton TM, Held H, Kriegler E, Hall JW, Lucht W, Rahmstorf S, Schellnhuber HJ (2008) Tipping elements in the earth's climate system. Proc Natl Acad Sci U S A 105(6):1786–1793
Levin SA (1992) The problem of pattern and scale in ecology. Ecology 73:1943–1967
Levin S (2010) Complex adaptive systems and the challenge of sustainability, In: SA Levin, WC Clark (eds) Toward a science of sustainability. CID Working Paper, n° 196:129–134. Harvard
Luginbuhl Y (1992) Nature, paysage, environnement, obscurs objets du désir de totalité. In: Robic M-C (ed) Du milieu à l'environnement. Economica, Paris, pp 12–56
Mancebo F (2010) Le développement durable, 2nd edn. Collection U, Armand Colin, Paris
Mollinga PP (2008) Boundary work an the complexity of natural resource management. Crop Sci 50(Suppl 1):S1–S9
Moret J (2004) La biodiversité à Paris. In: Michaud Y (ed) L'université de tous les savoirs. Odile Jacob, pp 217–238

Moser G, Weiss K (2003) Espaces de vie. Aspects de la relation homme-environnement. Armand Colin
Olander S, Johansson R, Niklasson B (2007) Aspects of stakeholder engagement in the property development process. In: Atkin B, Borgbrant J (eds) Proceedings of 4th Nordic Conference on Construction Economics and Organization, Research Report, n° 18. Lund University, Lund Sweden, pp 141–150
Ostrom E (1998) A behavioral approach to the rational choice theory of collective action. Am Polit Sci Rev 92(1):1–22
Pearce D, Markandya A, Barbier E B (1989) Blueprint for a green economy. Earthscan, Oxford
Proshansky HM, Fabian AK, Kaminoff R (1983) Place-identity: physical world socialization of the self. J Environ Psychol 3:57–83, Elsevier
Robinson J (2003) Squaring the circle? Some thoughts on the idea of sustainable development. Ecol Econ 48:369–384
Roseland M (1998) Towards sustainable communities: resources for citizens and their governments. New Society, Gabriola Island
Sen A (2009) The idea of justice. Allen Lane, London
Smith N (2002) New globalism, new urbanism: gentrification as global urban strategy. Antipode 34:427–450
Solow R (1974) The economics of resources or the resources of economics. Am Econ Rev 64:1–14
Stokke O (2009) The UN and development: from aid to cooperation. UN intellectual history project. Indiana University Press, Bloomington IN
Strayer DL, Beighley RE, Thompson LC, Brooks A, Nilsson C (2003) Effects of land cover on stream ecosystems: roles of empirical models and scaling issues. Ecosystems 6:407–423
Swart R, Robinson J, Cohen S (2003) Climate change and sustainable development: expanding the options. Clim Pol 3:19–40
Theys J, Emelianoff C (2001) Les contradictions de la ville durable. Le Débat 113:122–135
Turetsky MR, Harden JW, Friedli HR, Flannigan M, Payne N, Crock J, Radke L (2006) Wildfires threaten mercury stocks in northern soils. Geophys Res Lett 33:16
Turner MG (2010) A landscape perspective on sustainability Science. In: Levin SA, Clark WC (eds) Toward a science of sustainability. CID Working Paper, n° 196. Harvard, pp 123–129
Turner BL, Robbins I, Robbins P (2008) Land-change science and political ecology: similarities, differences and implications for sustainability science. Ann Rev Environ Resour 33:295–316
Turhnout E (2009) The effectiveness of boundary objects: the case of ecological indicators. Sci Public Policy 36(5):403–412
UN-HABITAT (2008) State of the World Cities Report 2008–2009 – Harmonious Cities
Van Noordwijk M, Tomich, Verbist (2001) Negotiation support models for integrated natural resource in tropical forest margins. Ecol Soc 5(2), article 21
Von Weizsäcker EU, Lovins AB, Lovins LH (1997) Factor four: doubling wealth, halving resource use. Earthscan, Oxford
WCED (1987) Our common future. Oxford University Press, Oxford
Weingart P (1983) Verwissenschaftlichung der Gesellschaft–Politisierung der Wissenschaft. Zeitschrift für Soziologie 12:225–241
Werling BP, Gratton C (2008) Influence of field margins and landscape context on ground beetle diversity in Wisconsin (USA) potato fields. Agric Ecosyst Environ 128:104–108
Wilbanks TJ (2003) Integrating climate change and sustainable development in a place-based context. Clim Pol 3:147–154
Winkelmann R (2012) Conspicuous consumption and happiness. J Econ Psychol 33(1):183–191

Part III
Some Governance Issues

The Legitimation of Global Energy Governance: A Normative Exploration

Sylvia Karlsson-Vinkhuyzen

Abstract Global energy governance has very limited legitimacy in the eyes of most governments. Although the concept has been starting to surface in academic papers it is still barely used in policy discussions. It is contested, almost taboo, to raise the need for international norms around energy production or consumption, although a significant step forward was taken by including energy as one of the Sustainable Development Goals proposed to the UN General Assembly. It is becoming a bit less contested to strengthen international collaboration on renewable energy and energy efficiency. Least controversial are efforts to collaborate around efforts to increase access to modern energy for those who are still deprived thereof.

In this paper I analyse in more detail the present lack of legitimacy of global energy governance and more importantly the possible avenues for strengthening it as this is a prerequisite for the fundamental dimension of society that energy production and consumption constitutes in a new social contract. I take as starting point theories of normative legitimacy that consider its two major components as being input and output legitimacy. I elaborate on the output related arguments to strengthen global energy governance – its role for building a sustainable global energy system and deep energy security – grounded in the principle of subsidiarity. I further explore the necessary elements to ensure input legitimacy of global energy governance relating to participation, transparency and accountability.

Finally I discuss the possible relationship between this normative analysis of the legitimacy of global energy governance and the subjective legitimacy of the same phenomenon among state and non-state actors. The latter is what matters in the negotiations to address energy not only in the Sustainable Development Goals but also in the climate regime.

Keywords Energy governance • International norms • Legitimacy • Sustainable development goals • Energy security

Sylvia Karlsson-Vinkhuyzen is an Assistant Professor at Wageningen University. Email. sylvia.karlsson-vinkhuyzen@wur.nl

S. Karlsson-Vinkhuyzen (✉)
Public Administration and Policy Group Wageningen University,
PO Box 8130, 6700EW Wageningen, The Netherlands
e-mail: sylvia.karlsson-vinkhuyzen@wur.nl

F. Mancebo, I. Sachs (eds.), *Transitions to Sustainability*,
DOI 10.1007/978-94-017-9532-6_10

Global energy governance has very limited legitimacy in the eyes of most governments. Although the concept has been starting to surface in academic papers it is still barely used in policy discussions. It is contested, almost taboo, to raise the need for international norms around energy production or consumption, although a significant step forward was taken by including energy as one of the Sustainable Development Goals proposed to the UN General Assembly. It is becoming a bit less contested to strengthen international collaboration on renewable energy and energy efficiency and least controversial are efforts to collaborate for increasing access to modern energy for those who are still deprived thereof (Karlsson-Vinkhuyzen 2010). A good sign of this is the inclusion, after much lobbying efforts, access to affordable, reliable, sustianable and modern energy for all as one of the suggested Sustainable Development Goals.

This chapter analyzes in more detail the present lack of legitimacy of global energy governance and more importantly the possible normative avenues for strengthening it. The rationale for this is that such strengthening can be seen as a prerequisite for the fundamental dimension of society that energy production and consumption constitutes in the new social contract that was discussed in the Third *Rencontres Internationales de Reims* on Sustainability Studies in June 2013. The starting point here, is the theories of normative legitimacy with two of their major components being input and output legitimacy. The chapter elaborates on the output related arguments to strengthen global energy governance—its role for building a sustainable global energy system and deep energy security—grounded in the principle of subsidiarity. Further, the necessary elements to ensure input legitimacy of global energy governance relating to its sources and process are discussed. Finally, conclusions include a brief discussion about the possible relationship between this normative analysis of the legitimacy of global energy governance and the subjective legitimacy of the same phenomenon among state and non-state actors. The latter is what matters in the negotiations to address energy not only in the Sustainable Development Goals but also in the climate change regime.

Energy in the New Social Contract

The way that we manage our relationship to the vast sources of energy this planet harbors is an essential component of a new social contract that could guide the development of our societies for the future. Our modern societies, and all the dimensions of these that have contributed to our increased well-being, security and development have been built on a strong addiction to cheap energy, mostly from fossil fuels (Smil 2003; GEA 2012). At the same time this addiction has, among many other things, enabled frightful advances in our ability to develop weapons to kill each other, made the air of our cities unhealthy to breathe and brought us climate change (GEA 2012, chapters 4 & 5). However, this addiction to fossil fuels has not been equally awarded all of humanity. Access to modern energy and thus the benefits of the services it provides has been and remains very unequal with 1.4 billion people having no access to electricity and 2.7 billion people who rely primarily on traditional biomass for cooking (GEA 2012, chapter 2). Ironically the poorest not only suffer considerable health and other consequences from being deprived of modern energy services (indoor air pollution etc.), they will in many cases also be the primary victims of the excessive use of fossil fuels by those who have access to them through impacts of climate change.

Based on these few facts it is reasonable to conclude that the whole global energy system, including the ways that energy is produced and consumed and the infrastructures that support, it has to dramatically change (Karlsson-Vinkhuyzen et al. 2012). Indeed it needs such a radical change that we cannot envision what it would look like (Des Bouvrie et al. 2013). The question that I raise in this chapter is what role *global* (rather than regional, national or local) energy governance could have in bringing about this change and how this role could be legitimized. Adopting a Sustainable Development Goal (SDG) on energy as proposed by the High-Level Panel of Eminent Persons on the Post-2015 Development Agenda (2013) would constitute a step towards strengthening such global energy governance.

A Meager History of Energy in the UN

Global governance of issues such as the environment and development has a long standing on the agenda of the UN. And although their governance may be insufficient and/or ineffective in a number of dimensions the role of international norms and organizations in these domains is seldom questioned. They are seen as having a legitimate role, that is their authority is seen as justified (Bodansky 1999). Indeed, many would like international norms and organizations to be strengthened (Biermann et al. 2012; Kaul et al. 2003). When it comes to global governance of energy the story is entirely different.

Global energy governance has had, and still has as will be described below, very limited legitimacy at least in the eyes of many governments (Bodansky 1999).[1] Furthermore, although the concept has started to surface in academic papers in the 2010s (Lesage et al. 2010; Karlsson-Vinkhuyzen 2010; Goldthau 2011; Van Der Graaf 2013) it is still barely used in policy discussions including those that de facto center on such governance for example in the advocacy for the Sustainable Energy Access for All decade and an energy related SDG. And if global energy governance has advocates in civil society they are neither visible nor vocal.

The illegitimacy of global energy governance in the eyes of most governments is strikingly manifested by its very humble presence on the agenda of the UN System since its inception. Energy has during close to the 70 years of UN history been subject to: a handful of scientific conferences or meetings, some committees mostly under UN's Economic and Social Council, a few intergovernmental negotiations

[1] I define global energy governance as encompassing those efforts that seek to address energy as a common affair in the international community (Karlsson-Vinkhuyzen 2010). This excludes possible governance measures among e.g. energy companies that focus on revenues for themselves.

over normative language and not the least considerable attention in the development lending or aid of the World Bank and some UN agencies (Karlsson-Vinkhuyzen 2010). Global governance on energy in the form of dedicated organizations, institutionalized cooperation or international norms has clearly been seen as an illegitimate sphere of UN action for much of the organization's history.

Nonetheless, there have been small steps in the direction of global energy governance in the first years of the 2000s. One the one hand energy in relation to sustainable development has been subject to negotiations of declarations and action plans in intergovernmental fora in the follow-up process the United Nations Conference on Environment and Development in Rio de Janeiro in 1992 (Najam and Cleveland 2003). This includes the meetings of the Commission on Sustainable Development that discussed energy in 2001 and 2006/7, the World Summit on Sustainable Development in 2002 and the United Nations Conference on Sustainable Development in 2012 (Rio+20). However, the texts that were adopted at these meetings were vague and without any clear role for global governance in achieving the adopted aspirations. The outcome documents of these meetings contained formulations such as:

> Governments, taking into account their national circumstances, are encouraged to: Develop and implement appropriate national, regional and international policies and measures to create an enabling environment for the development, utilization and distribution of renewable energy sources. (United Nations Commission on Sustainable Development 2001, Decision 9 /1, para 17a)

> We recognize that improving energy efficiency, increasing the share of renewable energy and cleaner and energy-efficient technologies are important for sustainable development, including in addressing climate change... We note the launching of the "Sustainable Energy for All" initiative by the Secretary-General... (United Nations General Assembly 2012, Para 128–129)

The initiative that governments did not endorse, encourage or support but merely 'note' – the UN Decade of Sustainable Energy for All 2014–2024 – is as most UN decades a very low key, bottom-up approach where governments decide what they want to work on and does thus not really indicate that the legitimacy of *global* energy governance has dramatically increased.

In line with the history of a very humble presence of energy in UN based global governance, energy was also glaringly absent in the Millennium Declaration and Millennium Development Goals (MDGs) (United Nations General Assembly 2000). In the non-governmental consultation processes towards a post-2015 development agenda and the SDGs there were efforts by many primarily UN agencies and non-state actors to include energy dimensions, both as one of the overarching targets (e.g. secure sustainable energy) and as being linked to a number of other goals and targets. For example, the High-Level Panel of Eminent Persons on the Post-2015 Development Agenda (2013) suggests a goal to "secure sustainable energy" and list the following examples of specific goals that could be included: double the share of renewable energy in the global energy mix; ensure universal access to modern energy services; double the global rate of improvement in energy efficiency in buildings, industry, agriculture and transport; and phase out inefficient fossil fuel subsidies that encourage wasteful consumption. Nilsson argues that "energy needs to be a key constituent of any globally agreed set of SDGs" that "the provision of energy services in poor economies should be an explicit goal" and that "SDGs should include goals for efficiency and practices and low carbon energy expansion" (Nilsson et al. 2012). Finally, the Global Thematic Consultation on Energy and the Post-2015 Development Agenda (2013) "call on all relevant actors to work together to develop and establish a global goal on energy" and conclude that there is "broad support for 'sustainable energy for all' as a global goal." The fact that these efforts were partially successful, as the proposed SDG no 7 is to "[e]nsure access to affordable, reliable, sustainable, and modern energy for all"[2] indicates that governments are becoming more comfortable with at least the aspirational goal-setting part of global governance for the domain of energy.

Legitimizing Global Energy Governance

At the heart of a social contract is what type of political authority has legitimacy, that is, what type of political authority is justified. It is not difficult to explain why governments do not consider global energy governance a legitimate activity. It is rooted in the close association of energy with national security; the state and its economic and military security was for most of the twentieth century at the center of concern and energy is a crucial element in both these dimensions of security (Peters 2004; Willrich 1976). Although the 1980s and 1990s saw developments that made energy to be seen more as a commodity of trade rather than an issue for security and geopolitics, the concern about energy security has for various reasons come back on the agenda in the early 2000s (Peters and Westphal 2013). Consequently, energy security is considered as a national public good with its provision often considered a priority for governments. Collaboration with other countries does not come easily within this paradigm and many win-win opportunities in energy investments, technology cooperation and governance are foregone (Karlsson-Vinkhuyzen et al. 2012).

The rest of this chapter will present alternative avenues for legitimizing global energy governance. I have earlier summarized the normative literature on sources of legitimacy in elements of international/global governance and used it to develop a framework for analyzing and comparing normative legitimacy (see Table 1). Sources of legitimacy in normative literature may of course not be identical to sources of subjective legitimacy – what is seen as legitimate by particular actors such as national governments (states). However, on the one hand there should be considerable overlap between sources of normative and subjective legitimacy (Black 2008), and on the other hand I would argue that also an elaboration of normative reasons for strengthening global energy governance is of value. For both these reasons I will use components of this framework when examining possible strategies to legitimize global energy governance and thus it becomes an exploration grounded in normative arguments.

[2] See http://sustainabledevelopment.un.org/focussdgs.html.

Table 1 Sources of legitimacy for elements of global governance

Components of legitimacy	Sub-components of legitimacy
Source-based legitimacy (input legitimacy)	Expertise
	Tradition
	Discourse
Process-based legitimacy (input legitimacy)	Government participation
	Non-governmental participation
	Transparency
	Accountability
Outcome-based legitimacy (output legitimacy)	Effectiveness
	Equity

Table 1 describes three different but very interrelated main components of legitimacy.[3] Each of these can in turn be divided into sub-components that enable a more detailed analysis.

Output Related Legitimation

The possibilities to legitimize stronger global energy governance related to its output naturally depend on what output is considered desirable. From a normative standpoint it is possible to formulate at least two encompassing desirable outputs of global energy governance. The first desirable output is a global sustainable energy system, which implies an energy system that is sustainable in environmental, social and economic dimensions over time. This means that the system of energy production and consumptions (and all the infrastructure and social institutions associated with it) would be one that for example minimizes the risk for dangerous climate change, reduces the vulnerability of economic development to high and fluctuating fossil fuel prices and makes energy sources cheaper and more accessible for future generations (Karlsson-Vinkhuyzen et al. 2012). An equity dimension of such a system would be sustainable energy access for all, a goal already adopted by the United Nations General Assembly through its endorsement of the Sustainable Energy for All Decade (see above). The second desirable output is the goal of achieving deep energy security. This is related to energy access but goes further. The concept of 'deep energy security' expands the traditional notion of energy security to encompassing human security; deep energy security is energy security that contributes to human security over space (from local to global) and time (that is, now and for future generations) (Karlsson-Vinkhuyzen and Jollands 2013). Deep energy

[3]Adapted from Karlsson-Vinkhuyzen and Vihma (2009), p. 410. The original framework was developed for the legitimacy of international norms, but I would argue that it can equally well be applied to other elements of global governance.

security is a necessary condition for human security and cannot be achieved unless the global energy system is sustainable.

Assuming that these two goals are worth pursuing the following question is then how we can identify if governance at the global level could support achieving these goals? Similar questions about allocating governance to higher levels have been asked in federal states but also very much in the evolving European Union. In the EU the principle of subsidiarity has been adopted as a guide to allocating governance between levels. The principle appears in two dimensions within the EU, a substantial dimension which is linked to input legitimacy that we will talk about later and a procedural dimension which is linked to finding the level of decision-making which is most effective (Føllesdal 1998). If we here focus on output based legitimacy it is the procedural part of the principle that becomes of interest. This principle in its EU interpretation comes to imply that:

1. Action should be taken at the level where it is most effective, the effectiveness condition, and
2. Action at the higher level should be taken when lower levels cannot achieve the adopted goals in isolation, the necessity condition (the latter may be result of either lower levels not having the capacity or not having the political will).

Applying this procedural dimension of the principle on energy asks for governance at the global level in two cases. Firstly, global governance is needed when it is effective. This can be the case for example in areas where action by individual countries or the market is not likely to be sufficient such as development of (accessible) knowledge and norms promoting sustainable energy or when it aims to strengthen the coherence of the international community's (intergovernmental organizations), support for sustainable energy. Second, global governance is needed when it is necessary. This can be the case when many countries such as Small Island Developing States do not have the capacity to build up renewable energy sources and when other countries may have the capacity but not the political will for promoting sustainable energy. Another factor that can necessitate global governance is when global institutions (either norms and/or organizations) are contributing to preserving a fossil fuel based unsustainable energy system. Here we can think of the policies of international financial institutions that still predominantly invest in fossil fuel based energy systems, rules on trade and intellectual property rights that may constrain widespread technology transfer or favor unsustainable investments.

In normative terms there seems to be strong legitimation possibilities for global energy governance related to its output.

Input Related Legitimation

The possibilities to legitimize global energy governance related to input can be explored along the sub-components of source based legitimacy; expertise, tradition and ideology on the one hand, and to the sub-components of process based legitimacy; participation, transparency and accountability on the other.

First looking at source based legitimation, it is clear that with so limited a history of addressing energy in global governance tradition is not going to do much for legitimation of UN based energy governance. However, it may play a role in the International Energy Agency's (IEA) efforts to be seen as *the* international organization on energy although it only has OECD countries as members. The IEA is also building much of its legitimacy on its energy expertise, as manifested in the annual production of the World Energy Outlook and its self-description as being "at the heart of global dialogue on energy, providing authoritative statistics, analysis and recommendations."[4] Ideologies that would be supportive of global energy governance could include those linked to world federalism, human security, fairness etc. Even adherents to liberalism could argue that market failures have to be addressed at the global level to manage these and to ensure a level playing field.

Moving on to process based legitimation this is linked to the substantive dimension of the subsidiarity principle that dictates decision-making as close as possible to citizens. This implies that these citizens should have some at least indirect access to the governance process through democratic institutions. If for effectiveness reasons we argue that governance is still needed at the highest, in this case global, level then the question instead becomes how to make governance at this level 'close' to the citizens – bringing some dimensions of democratic or similar characteristic elements that can give it democratic legitimacy. Possible sources of such legitimacy are the four sub-components of process based legitimacy outlined in Table 1.

The first sub-component of process-based legitimacy is governmental participation. Considering that at the moment perhaps the strongest intergovernmental organization on energy is the IEA whose membership is not universal this is a potential avenue for legitimation. The IEA does reach out to BRIC countries but it is a big step before it opens its doors to non-OECD countries as members and when it comes to governmental participation as a source of legitimacy it is decision-making power that counts. In contrast, UN agencies are mostly open for participation of all states but when energy is so low on the agenda they cannot do much. On the contrary, the International Renewable Energy Agency (IRENA) was established as a coalition of the willing outside the UN, they wanted to move faster among countries that had an interest in renewable energy.

The second sub-component of process-based legitimacy is non-governmental participation. Some political theorists have seen such participation as an avenue for strengthening the democratic character of global governance. However, strengthening the legitimacy of global energy governance through this avenue faces considerable challenges. There are very few international NGOs who act as advocates for strengthening global energy governance. There are a few working on energy access but there are hardly any voices raised for renewable energy and energy efficiency at least in the UN corridors when energy is discussed.[5] One reason for this could be

[4] See http://www.iea.org/aboutus/whatwedo/

[5] This is an observation from having followed the UN based negotiations on energy and some other international energy meetings in the 2000s. One example of an NGO advocating for energy access around the UN meetings is ENERGIA, an international network on gender and sustainable energy, see www.energia.org. This network is particularly interesting considering that energy is normally a very male dominated sector.

the technical nature of energy making it challenging for NGOs to take it up as a central topic.[6] Another reason can of course be that organizations that try to advocate for sustainable energy do not see any role for global governance in this. Perhaps a more fundamental question around participation is: How can local communities be engaged in a way that empowers them to identify their own goals and development pathways around energy in the context of a global SDG on energy?

The third sub-component of process-based legitimacy is transparency. Transparency is a major challenge in any global governance process. How can then a governance process on energy that on other levels is traditionally confined to small groups of closed networks be opened up and made transparent and accountable towards those whose lives their decisions influence?

Accountability is the fourth sub-component of process based legitimacy and it is a multidimensional concept. Its importance is emphasized by the High-Level Panel that argues that one of the five transformative shifts that should guide the post-2015 development agenda should be to "build peace and effective, open and accountable institutions for all" (High-Level Panel of Eminent Persons on the Post-2015 Development Agenda 2013). A prerequisite for holding actors to account is that there is transparency in who does what. In relation to international agreements such as SDGs the degree of implementation is an obvious activity that needs to be tracked through monitoring and reporting. However, countries are very reluctant to agree any monitoring by outside agencies, often claiming sovereignty reasons while probably well aware that monitoring is indeed what is needed to enforce norms. Having followed close-hand the fate of the proposal on reporting and follow-up of the energy agenda in the CSD (Karlsson-Vinkhuyzen 2010), where it was the issue that made it impossible for unwilling countries to accept the text, it is clear that any proposals around reporting have to be creative in approach to overcome this sensitivity.

In normative terms there are certainly a number of avenues to legitimize global energy governance related to its input although most of these would require a quite radical turn in global politics.

Conclusions

The analysis of normative components of legitimacy that could strengthen global energy governance shows that strong output based arguments can be made. Legitimation arguments based on the input to governance can also be made but seem to be quite challenging to realize. What matters in any efforts to actually strengthen global energy governance is, as I argued above, not normative but subjective legitimacy in the eyes of governments and other actors such as business and civil society. There is surely overlap between the components of normative legitimacy that I have just elaborated, and

(continued)

[6] Greenpeace is an exception here in the sense that they have even challenged the modeling efforts of the IEA and produced "their own" energy scenarios for the future with renewable energy playing a central role in those (Greenpeace International et al. 2012).

subjective legitimacy, but how much is open for research considering the deep rootedness of the national security paradigm. It seems clear that progress is only possible if this paradigm can be challenged by alternative discourses that could change the mind-set of leaders and governments. Such discourses, for example one on deep energy security, could not pick up without a switch in focus of national leaders towards considering impacts of their energy decisions on citizens of other countries and future generations. It is not enough to argue that it is necessary to base governance on national self-interest. The perspective has to expand towards global goods and benefits and from short to long-term horizons. Such a switch in mind-set and guiding value for decision making asks a lot of leaders – perhaps primarily political leaders but also individuals across organizations whether they have position of formal leadership or not. It really requires moral leaders in the sense where they have a "consistent orientation of service to the common good" (Anello 1997, p. 89) and a willingness to assume the personal risks inherent in dealing with resistance to change WHO (1988) quoted in Anello and Hernández (1996, ix). For an overview of a framework expanding on the concept of moral leadership see Vinkhuyzen and Karlsson-Vinkhuyzen (2013).

To conclude: If we adopted the goals of a global sustainable energy system and deep energy security and apply the procedural dimension of the subsidiarity principle we have some strong arguments for legitimizing stronger global energy governance in several areas if these goals are to be effectively achieved. We can also identify key aspects of strengthening the normative legitimacy of global energy governance through increased participation of countries and non-state stakeholders in the key institutions and processes, and a leap forward in opening up both the negotiation processes and their implementation to public scrutiny. At the same time I have illustrated how unlikely any strengthening of global energy governance is because of its low legitimacy in the eyes of many governments. The only way that I can see out of this deadlock is individual leaders with the courage to move into new territory beyond the institutional constraints that surround them.

References

Anello E (1997) Distance education and rural development: an experiment in training teachers as community development agents, PhD, University of Massachusetts Amherst

Anello E, Hernández JB (1996) Moral leadership, Núr University, Bolivia (Unpublished translation)

Biermann F, Abbott K, Andresen S, Bäckstrand K, Bernstein S, Betsill MM, Bulkeley H, Cashore B, Clapp J, Folke C, Gupta A, Gupta J, Haas PM, Jordan A, Kanie N, Kluvánková-Oravská T, Lebel L, Liverman D, Meadowcroft J, Mitchell RB, Newell P, Oberthür S, Olsson L, Pattberg P, Sánchez-Rodríguez R, Schroeder H, Underdal A, Vieira SC, Vogel C, Young OR, Brock A, Zondervan R (2012) Navigating the anthropocene: improving earth system governance. Science 335:1306–1307

Black J (2008) Constructing and contesting legitimacy and accountability in polycentric regulatory regimes. Regul Gov 2:137–164
Bodansky D (1999) The legitimacy of international governance: a coming challenge for international environment law? Am J Int Law 93:596–624
Des Bouvrie N, Karlsson-Vinkhuyzen S, Jollands N (2013) Responsibility for radical change in emission of greenhouse gasses. Paper presented at The Radical Emission Reduction Conference, Royal Society, London
Føllesdal A (1998) Survey article: subsidiarity. J Polit Philos 6:190–218
GEA (2012) Global Energy Assessment – toward a sustainable future. Cambridge University Press/International Institute for Applied Systems Analysis, Cambridge/New York/Laxenburg
Global Thematic Consultation on Energy and the Post-2015 Development Agenda (2013) Key messages, United Nations. Available: http://www.worldwewant2015.org/node/339192. Accessed 5 Dec 2013
Goldthau A (2011) Governing global energy: existing approaches and discourses. Curr Opin Environ Sustain 3:213–217
Greenpeace International, EREC and GWEC (2012) Energy [r]evolution. A sustainable world energy outlook, European Renewable Energy Council (EREC), Global Wind Energy Council (GWEC), Greenpeace International
High-Level Panel of Eminent Persons on the Post-2015 Development Agenda (2013) A new global partnership: eradicate poverty and transform economies through sustainable development. United Nations, New York
Karlsson-Vinkhuyzen SI (2010) The united nations and global energy governance: past challenges, future choices. Glob Chang Peace Secur 22:175–195
Karlsson-Vinkhuyzen SI, Vihma A (2009) Comparing the legitimacy and effectiveness of global hard and soft law: an analytical framework. Regul Gov 3(4):400–420
Karlsson-Vinkhuyzen SI, Jollands N (2013) Human security and energy security: a sustainable energy system as a public good. In: Dyer H, Trombetta MJ (eds) International handbook of energy security. Edward Elgar, Cheltenham, pp 507–525
Karlsson-Vinkhuyzen SI, Jollands N, Staudt L (2012) Global governance for sustainable energy: the contribution of a global public goods approach. Ecol Econ 83:11–18
Kaul I, Conceição P, Le Goulven K, Mendoza RU (2003) Providing global public goods. Managing globalization. Oxford University Press, Oxford
Lesage D, Van De Graaf T, Westphal K (2010) Global energy governance in a multipolar world. Ashgate, Farnham
Najam A, Cleveland CJ (2003) Energy and sustainable development at global environmental summits: an evolving agenda. Environ Dev Sustain 5:117–138
Nilsson M, Heaps C, Persson Å, Carson M, Pachauri S, Kok M, Olsson M, Rehman I, Schaeffer R, Wood D, Vuuren DV, Riahi K, Americano B, Mulugetta Y (2012) Energy for a shared development agenda: global scenarios and governance implications, Stockholm Environment Institute
Peters S (2004) Coercive western energy security strategies: 'resource wars' as a new threat to global security. Geopolitics 9:187–212
Peters S, Westphal K (2013) Global energy supply: scale, perception and the return to geopolitics. In: Dyer H, Trombetta MJ (eds) International handbook of energy security. Edward Elgar, Cheltenham, pp 92–113
Smil V (2003) Energy at the crossroads: global perspectives and uncertainties. The MIT Press, Cambridge, MA
United Nations Commission on Sustainable Development (2001) Commission on sustainable development. Report on the ninth session (5 May 2000 and 16–27 April 2001), Economic and Social Council, official records, Supplement n° 9. United Nations, New York
United Nations General Assembly (2000) Resolution adopted by the general assembly: 55/2, United Nations millennium declaration. United Nations, New York

United Nations General Assembly (2012) Resolution adopted by the general assembly, 66/288 the future we want. United Nations, New York
Van Der Graaf T (2013) Fragmentation in global energy governance: explaining the creation of Irena. Glob Environ Polit 13:14–33
Vinkhuyzen OM, Karlsson-Vinkhuyzen SI (2013) The role of moral leadership for sustainable consumption and production. J Clean Prod 63:102–113
Willrich M (1976) International energy issues and options. Annu Rev 1:743–772

From Government to Multi-stakeholder Governance for Sustainable Mobility

Marc Dijk

Abstract This chapter discusses an advanced form of stakeholder participation to enhance governance for sustainable car mobility. Our dynamic governance perspective is based on the idea that policies should be concerned not only with providing incentives and setting limits but also with providing orientation, stimulating mutual learning, fostering socio-technical alignment, making sure that a wide variety of options is explored, dealing with conflicting claims by technology actors and with learning about the effects of their policies. The perspective on governance of innovation that we take is an alternative to the dominating (static) welfare perspective of internalizing externalities.

Our chapter starts with an evaluation of role of policy and regulation in car mobility on Europe in the last 25 years. We find that policy instruments were mostly applied within a neo-classical economic welfare perspective based on the use of regulation and economic incentives. It has lead mostly to diffusion of technical fixes and incremental innovation of the dominating propulsion technology, internal combustion engines, not to a modal shift or shift to alternative propulsion systems. Attempts to foster electric vehicles have largely failed.

We discuss how an advanced form of stakeholder participation can enhance the effectiveness of governance for sustainable mobility. Although our discussion mostly concerns Europe, the application of our concept is transferrable to other continents and other sectors.

Keywords Multi-stakeholder governance • Sustainable car mobility • Policymaking • Collective action • Transport policies

The quest for a sustainable society challenges our political-institutional system. This chapter addresses the quest for sustainable car mobility and the implications

Marc Dijk is a Research Fellow at ICIS (International Centre for Integrated assessment and Sustainable development) of Maastricht University. Email. m.dijk@maastrichtuniversity.nl

M. Dijk (✉)
ICIS (International Centre for Integrated assessment and Sustainable development),
Maastricht University, P.O. Box 616, 6200 MD Maastricht, The Netherlands
e-mail: m.dijk@maastrichtuniversity.nl

F. Mancebo, I. Sachs (eds.), *Transitions to Sustainability*,
DOI 10.1007/978-94-017-9532-6_11

for governance. It argues that without an advanced form of stakeholder participation, sustainable mobility will be an impossible goal.

By subsector, road transport is the largest contributor to global warming (Fuglestvedt et al. 2008) and societies, in particular urban areas, have been burdened with a range of impacts from transport activities, such as fossil fuel consumption, harmful emissions, traffic accidents, congestion, noise, and fragmentation of scarce space (Adams 2005). Air quality near crowded roads remains problematic and reduces life expectancy in Europe by an average of almost 9 months (Krzyzanowski et al. 2005). Fossil fuels are becoming scarcer and accordingly become a source of economic and political instability. Without governmental action, the transport sector will increasingly contribute to climate change and other problems. Most governments have set goals to bring down emissions and fuel use, but environmental regulation on fuel economy of fossil-fuelled vehicles has not been enough to stop the growth of total carbon emissions and local air pollution.

Mobility issues are multi-faceted, since they include social, economic, and ecological as well as technological aspects. Policy concerns are typically interconnected and cross several policy fields. Societal stakeholders have conflicting perceptions of the problem, whereas gains and burdens of transport activities are distributed unevenly over various societal groups. Transport policies have been partially successful in the last 20–30 years. The increase of safety levels and decrease of polluting gases in Europe (such as particulate matter, nitrogen oxides) can be attributed directly to transport-related policies. However, greenhouse gas emissions and (related) fossil fuel demand remain growing.

Public transport has been developed, but never attracted a major market share. Given the inevitable future growth of vehicle-kilometers worldwide, the goal to bring down carbon emissions implies either decreasing the emission per vehicle-kilometer significantly, or replacing car kilometers with lower emission alternatives (train, bus, bike etc.). Since the 1970s, especially electric and hydrogen cars have received much attention time and time again as the "zero-emission solution"- but a significant market has never been established for either of them. All in all, we find sustainable car mobility remaining a major societal challenge.

This chapter discusses an advanced form of stakeholder participation: periodic stakeholder arena's to enhance the effectiveness of governance for sustainable car mobility. Our dynamic governance perspective is based on the idea that policies should be concerned not only with providing incentives and setting limits but also with socio-technical alignment, providing orientation, stimulating mutual learning, making sure that a wide variety of options is explored, dealing with conflicting claims by technology actors and with learning about the effects of their policies. The perspective on governance that we take is an alternative to the dominating (static) welfare perspective of internalizing externalities.

Governance was traditionally associated with government and governing: to rule or control with authority, but it has emerged into a new meaning which is much broader than government. Governance does not point to state actors and institutions as the only relevant ones, but focus on the role of networks consisting of actors from different societal domains in the pursuit of common goals. From this perspective, governance can be seen as attempts to understand the changing patterns of state-societal interaction. It can be seen as a complex multi-actor process, involving representatives of the government, the business community and the civil society, allowing a plurality of values, beliefs, needs and interests to be merged into collective action. Governance can also be defined as steering or co-ordination in networks to achieve a common goal. However, this societal steering in the context of governance differs fundamentally from classic policy making and implementation.

Our chapter starts with an evaluation of role of policy and regulation in car mobility on Europe in the last 25 years. We find it has lead mostly to incremental innovation of the dominating propulsion technology, internal combustion engines, not to a transport modal shift or shift to alternative engines.

We discuss how an advanced form of stakeholder participation can enhance the effectiveness of governance for sustainable mobility. Although our discussion mostly concerns Europe, the application of our concept is transferrable to other continents and other sectors.

A Government Failure in Sustainable Car Mobility

Policies to reduce total carbon emissions from car mobility have been ineffective. An integrative innovation perspective helps to understand why. It shows how various interaction effects provide stability to established practices.

Environmental policymakers in Europe have introduced various instruments to mitigate the (total) CO2 emissions of car use. Over the last two decades technological solutions have received most attention, and electric vehicles, hydrogen vehicles, hybrids, clean and small diesel and gasoline vehicles have all been in the spotlights as possible low emission vehicles (LUVs) and ultra low emission vehicles (ULEVs). There has been considerable debate on which technology is most promising. This is a politicized debate, influenced by technological uncertainty, vested economic interests, and ecological objectives, between various member states in Europe (Jacob et al. 2005). In the remainder of this section we discuss four policy instruments that were applied in Europe concerning CO2 emissions of car mobility: emission regulation, voluntary agreement, R & D subsidies, and consumer tax exemption schemes (as identified in Dijk and Kemp (2012)).

The first instrument, the stepwise emission standard scheme (Euro 1–5) triggers firms to focus (only) a few years ahead. They do not provide an incentive to go beyond the standards, showing the limitation of standards. From a business perspective, it is not very economical to invest much in a new, still immature technology (Dijk and Montalvo 2009), since those efforts will lead to losses in the short term. Competition on the present market is fierce and it is found (relatively) more urgent to invest in the incremental innovation of the existing technology. The emission schemes do not compensate for the (market) punishment of firms with a longer time focus. Firms that invest large portions of their R & D portfolio in future engines (for launches 5–10 years ahead) draw from their engineering budgets of forthcoming engines, pruning their competitiveness in the next few years, and running the risk of being outcompeted.

We found that the second instrument, the voluntary covenant, was only slightly effective. This too is in correspondence with earlier studies (Klok 1989; OECD 2003). The progress that was made came from wider application of existing technology: direct injection systems. The third instruments, tax exemption schemes at the user side have been applied on national levels. Tax benefits for hybrid-electric vehicles have stimulated their sales in various countries, although these have not exceeded market shares of a few percent. To what extent has this triggered car firms to develop more hybrid electric vehicles? This effect has not been substantial, sine in the decade following the introduction of Toyota's Prius (worldwide in 2000), only two car firms, Honda and Toyota (including Lexus), had hybrid models on the European market (recently there are a few more). Although the tax benefit also counted for electric vehicles, it has not triggered established car firms to launch electric models. There is only one firm offering electric vehicles in most European markets, the Norwegian Think, which is a new market entrant.

The fourth instrument applied, R & D subsidy programs, is of special interest, since it has traits of a dynamic governance approach. The European policies on electric and hydrogen vehicle technology, such as support of R & D via the Hydrogen and Fuel Cell Technology Platform (HFP) and vehicle demonstration projects contributed to improved versions of electric and hydrogen vehicles, and provided experiences with actual usage of these vehicles in practical settings, which otherwise probably would not have been performed. Demonstration projects foster learning between users, producers and infrastructure at an early stage, with lessons feeding into private and public decision-making. In the same vein, the establishment of a platform is important because it structures the communication between various stakeholders in the sector: firms, research centers, universities, regulators, NGO's, consumer groups. However, demonstration projects mostly showed that electric and hydrogen vehicles have not yet reached the necessary maturity for broader market applications. The average willingness of car firms to further develop hydrogen vehicles is low (Dijk and Montalvo 2009). Therefore, this R & D instrument has not yet resulted in car firms launching this type of vehicles, and it's not clear whether the platform has made market application more likely than before. The development of electric and hydrogen technology took place alongside further improvements of ICE technology. The various technological trajectories are as moving targets and interact with each other, mostly in a competitive relation. The isolated character of the R & D subsidies, disconnected from regulation in the ICE regime, is a disadvantage of this measure.

What was the effectiveness of the policy instruments in terms of market shares of cleaner vehicles in Europe? The shifting market shares of new, established and refined propulsion technologies in the last 20 years can be summarized in the following four phases (see Fig. 1). In the first phase, until about 1995, we found incremental innovation of the regime technology, internal combustion engines, most importantly through carburetors being replaced by electronic injection systems. The second phase is an odd intermezzo of electric vehicles being launched on the car market by most large manufacturers, triggered by regulation in a few American states. This meant that the market diversified, since product launches of a non-regime

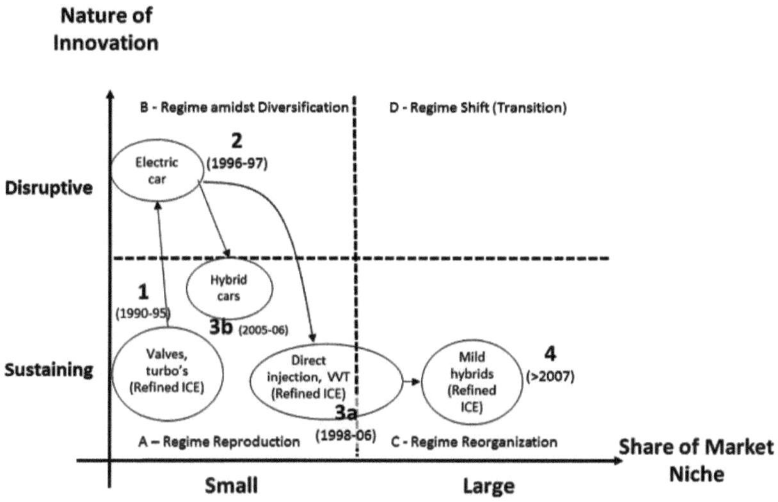

Fig. 1 Four phases in the evolution of the car engine market after 1990 (Dijk 2014)

technology were growing faster than that of the regime. After 1997 however, a phase (three) started, where the growth of electric models stopped and most models were even withdrawn from the market. By contrast, the conventional diesel engine went through a step of improvement when direct injection systems became widely applied and, later on, also particle filters were added. The next phase, from about 2004–2006, is when hybrid-electric vehicles catch momentum with a few products launches and sales growing with double digits. However, after 2007 it shows that apart from Toyota and Honda, other firms keep postponing their plans to launch hybrid versions. Instead, firms focus on launching so-called *cleantech* versions of internal combustion engines, which include electronic start-stop systems and special transmission software, with Volkswagen's *Bluemotion* series being one of the early movers.

Throughout this overview of 20-year history we are struck by the stability of the established regime. There is a tendency to avoid radical innovation (i.e. regime disruption) or even transformation, but instead an inclination to incremental innovation. This pattern has been observed earlier in the succession from sailing ships to steamships (and therefore it has been referred to as "the sailing ship effect" (Ward 1967). The pattern results, primarily, from defensive strategies of incumbent firms, who find hybrid solutions a less risky and, therefore, more attractive strategy.

All in all we find that the four instruments in Europe have led to diffusion of technologies and incremental innovation within the IC trajectory, not to a disruptive shift to alternative mobility options. This confirms the proposition of the typical response of industries to environmental policy instruments (Foxon and Kemp 2006).

The problem for policy (instruments) is that there is not one innovation that needs to be promoted, but a mix of entangled innovation pathways that interact. The societal attention for electric vehicles (EV) as a potentially green-car solution emerged amidst various possible green alternatives (hydrogen cars, biofuels, etc., but also shift to public transport or intermodality)[1] and amidst a well-established regime (gasoline, diesel car mobility). There are elements of symbiosis, competition and paralysis between green passenger mobility options– not only in a sense of technology, but also regarding their social and user context.

Therefore, policy instruments for eco-innovation, such as sustainable mobility, are surrounded by significant uncertainties, including the unpredictability of duration of the policy instrument itself. The uncertain nature of this type of innovation can deliver perverse consequences, an example outside Europe being the Zero Emission Mandate, a radical regulation introduced by the Californian government which stipulated a growing market share of zero-emission vehicles after 1996, but found only hybrid-electric vehicles (HEV) (a petrol car with a short electric range) being successful 10 years later. Ironically, HEVs benefitted from the progress on disruptive EVs in the 1990s and helped the ICE-regime to sustain. EVs, ICEs, HEVs and fuel-cell vehicles are typically treated as independent entities, whereas in practice they partly benefit from each other.

From Government to Governance

In today's more interconnected world, a wider range of different actors often influences decisions. Non-governmental organizations (NGOs), corporate and financial interests, and consumers 'and citizens' groups have won a louder voice in influencing decisions. The car mobility policies in the EU that we discussed were based on some level of stakeholder consultation and involved some form of stakeholder platform. Nevertheless, the overall policy framework was still fairly classic policymaking with the *government* as the dominant actor. We have discussed how this approach was ineffective in promoting sustainable mobility in terms of reducing total carbon emissions.

New governance approaches, such as interactive arrangements, networks and partnerships, have emerged in some societal sectors and countries. Here we discuss and plea for a periodic stakeholder arena for the transportation sector. The idea of stakeholders' arena is drawn from studies of transition management (Loorbach and Rotmans 2006; Loorbach 2007), and is rooted in fields such as multi-level governance, interactive policy-making and adaptive management. The concept of sectoral stakeholder arena's assumes that, although the quest for a sustainable society cannot be managed in terms of command and control, it can be managed in terms of influencing and adjusting: a more subtle, evolutionary way of influencing the direction and pace of developments in the sector. A further assumption is that structural and institutional changes, or transitions, in societal sectors as transport (or agriculture, health care etc.), can be triggered when the stakeholder arena focuses on the following role and activities:

[1] Smoother links between cars, public transport, bikes etc.

- The establishment and development of a multi-stakeholder innovation network, consisting of a diversity of actors: local, national and supra-national policymakers, car manufacturing businesses and first tier suppliers, scientists (engineers, marketing, behavioral and policy scientists), consumer organizations, NGO's etc., with an independent, "technical" chair.
- Problem-structuring especially in the early phase of the arena, much attention would be focused at highlighting the plurality of perspectives on the issue: on definitions of what sustainable, low-carbon mobility actually is.
- Visioning collective generation of long-term integrated future visions and backcasting activities (scenario development).
- Implementation and execution of practical experiments.
- Systematic monitoring, evaluation and adaptation of process and content.

The periodic stakeholder arena for the transportation sector would meet for instance twice a year (e.g. in spring and fall). The joint search- and learning process that the multi-stakeholder engaged in operates on a distance from the political arena (arena of current policy), in advisory role. The learning process has three components: learning-by-doing (developing theoretical knowledge and testing that by practical experience), doing-by-learning (developing empirical knowledge and testing that against the theory) and learning-by-learning (developing learning strategies, applying and evaluating them).

The concept of a periodic stakeholder arena links monitoring innovation dynamics with innovation governance. Monitoring innovation dynamics is focused on recognizing emergent ideas and trends throughout the whole spectrum of social order mechanisms, such as markets, networks, institutions, policies, individual behavior and autonomous trends. Sectoral innovation governance aims at steering those developments that can be influenced towards a direction in which those developments reinforce each other. The basic idea is that a long-term vision is formulated in a participatory process. From there, several sub-themes can be distinguished for which target images can be formulated, including pathways that lead to these future images. These images and pathways should be developed and explored through the use of scenarios, risk- and uncertainty-assessments. The next step is actively spreading the ideas to their home base or other networks that together form an innovation network. The innovation network stimulates the next step towards conducting practical experiments in experimental playgrounds on the practical level. The periodic stakeholder arena operates on a strategic level, focusing on the development of visions, new paradigms and pathways. But through the stakeholders it also affects operational level where experiments are carried out (together or alone) and the innovations are materialized. In order to arrange this at least two styles of governance need to go together (Loorbach 2004). At the strategic level a style of facilitating, exploring and enabling is required to settle the arena, choosing who participates and facilitating the long-term visions. At the more practical level, individual or collective actions of stakeholders need to be managed to stimulate the generation and transfer of ideas and knowledge. The two roles or activities of the arena need to interact in order to adjust to and learn from each other (learning by doing, doing by learning), top-down and bottom-up. Accordingly, the number of joining actors may increase over time, creating a growing network that has formulated a coherent sustainability perspective and is consistent (modulates) at strategic and operational level. Transition management as a whole is an advanced form of multi-level governance that requires different forms of governance at each level.

The European transport sector is of course a very large sector, and therefore one stakeholder arena may not be the right approach, but instead a range of arenas, distributed by region or country and also type of mobility, with one coordinating body. Here the Dutch transition approach (2008–2012) with the seven transition platforms for sustainable energy may be an example of interest (Kemp 2009). The platforms served as vehicles for learning and action for innovative solutions identified by business and experts (micro co-generation, battery electric vehicles, energy producing greenhouses, and others). Through the platforms the interest in society (business) in innovative change was mobilized. Strategic issues were considered via the Coordination-body Dutch Energy Transition created in 2008. The Coordination-body is responsible for developing an overall vision for the energy supply (electricity and heat) in the Netherlands and to formulate a strategic agenda based on inputs of the platforms. The whole approach is set up as a vehicle for sociotechnical change *and* policy change in a coordinated manner. The Dutch transition approach seeks to encourage industry to work on low-carbon innovations including those that are not yet ready for the market, and to assist in the development of those innovations. Generic policies are constructively combined with technology-specific policies, although a weakness of the Dutch approach was it was used as a push approach whereas it should have been used as a push and pull approach, by also using regulation. This (national) approach could be a prototype for a European platform for sustainable car mobility.

Conclusion

The progression of societal change towards sustainability is narrowly interconnected with the political-institutional context of society, where we see the need for a further shift from government (and its state-centric approach) to governance (and its more pluralistic approaches). The nature of sustainability issues – typically multi-faceted, involving social, economic, ecological as well as technological aspects, and with stakeholders have conflicting perspectives on the problem, and gains and burdens of current activities distributed unevenly over various societal groups – requires a more dynamic governance approach. The periodic multi-stakeholder arena that we have discussed for sustainable mobility enables the governance process to be:

- Mindful to interaction effects, such as societal innovation and the effect of policy instruments in inevitably surrounded by uncertainty and unintended consequences. In the arena these effects are explicitly discussed.

(continued)

- Continuous dialogue between stakeholders; the recurrent and scheduled character of the arena support commitment between stakeholders, which can trigger collective action,
- Scenario-based policies, since short-term policies are always discussed in relation to longer-term visions.
- Co-production of knowledge; scientific knowledge (engineering, marketing, behavioral, policy etc.) is combined with stakeholder knowledge (including policymakers). This prevents scientific solutions turning out to be impossible at a late stage, and also stakeholder's discussions losing connection with scientific studies.

In other words, the arena is a place or vehicle for both socio-techno-economic and policy learning.

Two questions may emerge: Who is invited for the arena? The arena should be open to any relevant stakeholder as long as one supports the quest for a sustainable mobility system. What is the legitimacy of the sectoral stakeholder arena as we have described it? In our view the arena has an advisory role and can therefore operate next to the existing legitimate, decision-making structures.

The Dutch government can be an example here, with a scheduled spring and fall deliberation with employer and employee representatives regarding wages and loan taxes etc. These meetings don't have decision-making power in themselves – the ministers of economic and social affairs have the decision-making power and remain responsible – but they take this consultation very seriously and also engage in agreements with them. For these agreement counts again: the ministers of economic and social affairs remain responsible for decisions taken (executive power) and the parliament the controlling (and legislative) power.

It can be debated whether the time frame for success is short enough for the timing of sustainability challenges. Also, there is a threat that powerful incumbents stakeholders dominate the arena. These are valid points for concerns (Ashford and Hall 2011) in connection to transition management), and deserve serious attention from the arena chair and members.

Public and stakeholder participation lies at the heart of the democratic process and has been an important part of decision making for millennia, more widespread over the last centuries. Rousseau's idea of the social contract between the governed and the government has since become the cornerstone of many political philosophies of government. In this chapter we have argued for an innovative form of stakeholder participation, which implies a slight adaptation of how the contract is currently functioning in most societies. We have argued that classical policymaking and one-directional implementation (based on limited stakeholder consultation) is too vulnerable for unforeseen interaction effects or strategic circumvention and waiting games.

(continued)

The complexity of sustainable development requires a stakeholder dialogue, not just bilateral between the government and a stakeholder, but an arena of stakeholders of the sector, in which interaction patterns in the sector and the perspectives of stakeholders become explicit. This cannot a once-in-a lifetime meeting, but it needs to be a periodic, e.g. twice a year, event, in order to keep aligning expectations, goals, and subsequent actor actions (new policies / adaptations, investments, consumer behavior, etc.). The complex nature of sustainability requires mutual, collective action and endeavor.

References

Adams J (2005) Hypermobility: a challenge to governance. In: Lyall C, Tait J (eds) New modes of governance: developing an integrated policy approach to science, technology, risk and the environment. Ashgate, Aldershot

Ashford N, Hall R (2011) Technology, globalization, and sustainable development. Yale University Press, New Haven

Dijk M (2014) A socio-technical perspective on the electrification of the automobile: niche and regime interaction. Int J Automot Technol Manage 14(2):158–171

Dijk M, Kemp R (2012) Evolutionary policy options for steering transition to low-carbon cars in Europe. In: Nilsson M et al (eds) Paving the road to sustainable transport: *Governance and innovation in low-carbon vehicles*. Routledge, New York

Dijk M, Montalvo C (2009) Ultra low emission vehicle development: a study on the drivers for car firms and the implications for sustainable development policy. Int J Automot Technol Manag 9(2):191–208

Foxon T, Kemp R (2006) Innovation impacts of environmental policies. In: Marinova D (ed) The international handbook on environmental technology management. Edward Elgar, Cheltenham

Fuglestvedt J, Berntsen T, Myhre G, Rypdal K, Bieltvedt Skeie R (2008) Climate forcing from the transport sectors. Proc Natl Acad Sci U S A 105(2):454–458

Jacob K, Beise M, Blazejczak J, Edler D, Haum R, Jänicke M, Loew T, Petschow U, Rennings K (2005) Lead markets for environmental innovations. Physica-Verlag, Heidelberg/New York

Kemp R (2009) The Dutch energy transition approach. IEEP 7:291–316

Klok PJ (1989) *Covenanten als instrument van milieubeleid* (Covenants as an instrument in environmental policy), unpublished PhD, Twente University, Enschede

Krzyzanowski M, Kuna-Dibbert B, Schneider J (eds) (2005) Health effects of transport related air pollution. World Health Organization, Copenhagen

Loorbach D (2004) Governance and transitions: a multi-level policy-framework based on complex systems thinking. paper Readen at Conference on Human Dimensions of Global Environmental Change, Berlin

Loorbach D (2007) Transition management: new mode of governance for sustainable development. International Books, Utrecht

Loorbach D, Rotmans J (2006) Managing transitions for sustainable development. In: Olshoorn X, Wieczorek AJ (eds) Understanding industrial transformation. Views from different disciplines. Springer, Dordrecht

OECD (2003) Voluntary approaches for environmental policy: effectiveness, efficiency and usage in policy mixes, report. OECD, Paris

Ward WH (1967) The sailing ship effect. Bull Inst Phys Phys Soc 18:169

Territorial Resources and Sustainability: Analyzing Development in a "Post-Fordist" Scenario

Bernard Pecqueur and Paulo Freire Vieira

Abstract In response to the challenges imposed by the effects of economic and cultural homogenization and deterritorialization brought on by Fordist rules of economic growth, territorial development strategies cannot be separated from the quest for sustainability - in particular that of the so-called "territorial resources". But sustainable development is usually seen as a global response to global issues, whereas territorial development usually takes place at an infra-regional scale, mobilizing local actors concerned by local issues. Our purpose then is to show that it is the confrontation of two systems of reasoning which enables the issues of a "post-Fordist model" to be apprehended. In this sense, we examine what could be termed a flexible and adaptable model of "sustainable territorial development", whose roots are anchored in the tradition of "eco-development" research.

Keywords Territorial resources • Post-Fordism • Bottom-up development • Systems approach • Territorial economics

The crisis of the dominant "Fordist" production model can be seen in the end of the domination of the industrial paradigm as the essential way of describing the value-creation process, and above all in the end of a production model based on individualized productivity and the generalized production of standard objects, i.e. ones which can be endlessly and identically reproduced. Nevertheless, the era now beginning, called in the absence of a better term "post-Fordism," does not

---

Bernard Pecqueur is a Professor of Economic Geography at the Université Joseph Fourier of Grenoble. Email. pecqueur@ujf-grenoble.fr

Paolo Freire Vieira is a Professor of Sociology Universidade Federal de Santa Catarina.

B. Pecqueur (✉)
UMR PACTE CNRS, Université Joseph Fourier Grenoble, 14 bis avenue Marie Reynoard, 38000 Grenoble, France
e-mail: pecqueur@ujf-grenoble.fr

P.F. Vieira
Universidade Federal de Santa Catarina,
Campus Universitário Reitor João David Ferreira Lima Trindade,
Florianópolis, Santa Catarina, Brazil

F. Mancebo, I. Sachs (eds.), *Transitions to Sustainability*,
DOI 10.1007/978-94-017-9532-6_12

seem to be solving the relationship to resources which has held sway since the Industrial Revolution. This is obvious when we face the simultaneous crises of globalization, production, finance and the management of the resilience of socio-ecological systems.

Current French-language literature, in which regulation theory (Boyer 2004) had previously developed the concept of 'Fordism' as a specific accumulation system, is now beginning to look closely at what will come after, through the analytical lens of cognitive capitalism (Colletis and Paulré 2008) and from a geographical perspective: "We are not only witnessing the redistribution of production sites on a global scale, but also the emergence of new organizational forms of production, with the emergence of supply and logistics chains." From this, the existence of clusters and of all the related types of specialized productive organization has been postulated (Becattini et al. 2009); more generally, it is now being assumed that regional and local scales are gateways to understanding current conditions of globalization. In the words of J. L. Klein (2008): "The local could be a basis for post-Fordist institutional reconstruction."

What, however, are the basic elements of such reconstruction?

The prevailing context is currently that of globalization, from which arises the first paradox we will examine: territorial construction processes as undertaken by local actors can be construed as ways for local economies to adapt to the effects of globalization as a spatial indicator. The second paradox is that territorial dynamics cannot be separated from sustainability (in particular that of territorial resources). Sustainable development is usually seen as a global response to global issues – one taking into account the complex relationships between the dinamics at the biosferic level and the specific features of regional and local dynamics – whereas territorial development takes place at an infra-regional scale and with groups and "communities" concerned by local issues. As has already been stated: "In view of the overall scheme of things, it seems incongruous to deal with sustainable development in local territorial contexts, since it would entertain the idea that sustainable development could be envisaged in an incomplete, itemized way" (Pecqueur and Zuindeau 2010, p. 49).

Our purpose here is therefore to show that it is the confrontation of these two systems of reasoning which enables the issues of the post-Fordist world to be apprehended. On the one hand there is a global, systemic view of the integrated, cooperative management of natural resources, space and the quality of life (sustainable development), and on the other a partial, specific view of the processes of creating and sustaining resources (territorial development).

The present chapter will first identify the most important aspects of what post-Fordism could be like when influenced by the territorial development approach, suggesting either a new geography of capitalism (Bouba-Olga 2006), or a shaking up of "reductionist notions of development and globalization culminating with the TINA philosophy" (Zaoual 2005, p. 15). We take as a starting point the characteristics going beyond the central core of the Fordist model: the relegation of material objects, the disconnection between places of production and consumption as superimposed areas, the emergence of a meso-economic scale and an awareness of the finite nature of productive resources and the pressing need for economic, social and environmental considerations to be organically integrated – as equals and over a long-term perspective.

We seek to validate the hypothesis of how the characteristics of the Fordist world are merging into the foundations of a post-Fordist model.

In the second part, we aim to highlight the opposing characteristics connecting two types of currently emerging development patterns. On the one hand, there is the pattern of territorial development, characterized by specificity and incompleteness, and on the other hand sustainable development, seen as an approach dealing with the interrelations between the socio-economic, the socio-cultural, the socio-political and the socioecological spheres - at the local, regional, national and international scales. The two processes and the complex and controversial questions they raise are confronted in the third part, where we examine what could be termed a "sustainable territorial development" model.

Some Radical Changes Emerging with Post-Fordism

Relegated Material Objects

The public generally considers a mechanical product to be more reassuring for politicians and others concerned by employment issues than a tourist service, and certain types of production seem more "proper" than others. This shows the extent to which Fordism is identified with the material aspects of production.

Material objects are at the core of Fordist functioning in the sense that the latter gives rise to specific technical systems having cumulative effects on consumption.[1]

Secondly, material production makes it easier to calculate individual productivity, the basis for calculating economic performance. In the words of Veltz (2009): "The number of pieces produced per day per worker can easily be counted, but how can the ability to react to interpersonal situations be measured … or the ability to build meaningful cooperation … performance has become a composite notion with multiple dimensions."

Finally, consumption itself is evolving towards the non-distinction of goods and services. In the words of Moati (2001), the consumer is thus perceived as a "utility-producing micro-firm" not so much seeking to acquire goods as to find "solutions to problems of consumption" in the shape of "consumption bundles."

One can thus speculate on what constitutes "real" production – probably a mix of several products and services where the specific, technological cognitive content (Moulier-Boutang 2007) is growing.

---

[1]We are referring here to the concept of 'technical system' described by Gille (1978) in his work *Histoire des techniques*, where the author notably shows how the steam engine, the British Industrial Revolution's key invention, brought about the development of maritime and rail transport, as well as the textile boom and progress in materials (iron, cast iron, steel), building, machine-tools, etc.

The Disjunction Between Places of Consumption and Production

Fordism attaches the worker to his or her place of production. There should be no wide spaces or long distances between workplace and dwelling place. Factory and residence are in close proximity. The overwhelming success of the system resides in the employee taking on the status of consumer on the spot.

The representation of an economic territory is thus still that of a micro-nation with the accounting equality of the national model in which, in the case under consideration, everything produced gives rise to an income, which is disposed of completely, either on consumption or in savings. Territories are thus small productive systems in which other actors – especially consumers – are accorded scarcely any room. The more time passes and as ever-smaller scales come into the observer's focus, the less places of production, income and expenditure seem pertinent. In other words, what is good for a company is a priori no longer automatically good for the inhabitants who host it.

Territories thus seem to be relatively autonomous, with a proliferation of initiatives and dodges for getting through the economic crisis. It is a growing tendency, which will profoundly modify actors' strategies and call into question the interlocking, traditional scales of public policy (Europe, nations and regions).

Such trends have been analyzed in studies dealing with "on the spot" economies, especially those focusing on future consequences of the disjunction dealt with here (Talandier and Davezies 2009).

The End of Macroeconomics?

Individual delocalization as a tendency of globalization tends to illustrate the typically Fordist mechanism of a quest for low production costs, continually separating producer communities from consumer communities. Current trading entities are less and less Nation-States and increasingly "regions," as the word is used in English, or "territories" (French territories) in the sense that the word is being used in Europe, i.e. constructed (rather than given) spatial structures. This calls into question Ricardo's model of international trade based on "comparative advantage".

Representing the world in terms of several interconnected "wholes" no longer works.

It follows that the easier it is for actors to accede to every corner of the world, the more they require the mediation afforded by "territories"

It is therefore our hypothesis here that, by the yardstick of globalization, comparative advantage turns into a "differentiating advantage." Territories should not therefore try to become specialized within a comparative framework, but rather, when the rules of competition are impossible to respect, get around them by concentrating on producing things they have – ideally speaking – a monopoly of. Such action would call into question the system of division based on national production (given areas), since initiatives would not come from policies centered on a Nation-State but from groups of actors rallying around to solve a production problem. Such groups exist at an intermediate scale between the individual and the "whole," a scale we call the "meso-economic."

If the tendency is confirmed, "multi-scalar" world and complex spatial combinations will emerge and impose themselves on actors, completely transforming global/local relationships (Vanier 2008).

Becoming Aware of the Finite Nature of Productive Resources

One of the basic postulates of the Fordist model was the idea that resources were infinite and that even if they ran out they could still be exploited by finding substitutes (nuclear power was one such dream). Not long ago, a well-known economists' society (Le cercle des économistes 2009) reacted to the question of resource depletion with the reminder that the management of scarcity is precisely the goal of the economic science. As they put it, "our ambition as economists is to combat scarcity with absolutely no intention of calling growth into question." Thus continues the dream of endlessly drawing on supposedly infinite resources. We are not trying to predict the long-term future, where a large number of substitutes will certainly come into use, but as the century marches onward there are definitely questions to be raised about the finiteness of productive resources. Bourg and Whiteside (2010) have suggested an "anthropology of the finite and the infinite" (p. 25 et seq.) where they first of all distinguish the Age of the Ancients, during which "condemning the majority to slavery seemed to be the condition for fulfilling the humanity of a small number." Then came modern democracy "opening up human aspiration to the infinite, as a technical action." Fordism is a true successor to the dream of overcoming and pushing back the limits of scarcity.[2] However, an awareness of the end of the uncontrolled, unmanaged abundance of nature's resources is finally dawning upon us. "We are increasingly coming up against multiple limits. Our feeling of finiteness is thus very different from that of the Ancients, stemming, as it does not from a form of a priori wisdom, but from the multifarious failure of the modern assertion of infinity. This finiteness concerns firstly our energy, mineral and fresh-water resources and, finally, the biotic ones." (Bourg and Whiteside 2010, p. 29).

Such trends are having clear impacts, not only on the spatial structuring of national territories, but also on traditional State structures, as shown by Saskia Sassen (2006). Sassen clearly shows that "territory" as an analytical category is indeed emerging as a means of assembly and as a key to analyzing globalization.

[2] Bourg and Whiteside (2010) evoke the economist Robert Solow, who defends the idea that the destruction of some received natural capital by a previous generation does not harm the following one.

The emergence of a post-Fordist model can be sensed in all of this, one dealing critically with current economic and cultural globalization trends by creating new relationships between the local and the global, new cognitive processes involving territories (in the plural) and a new geography of natural resources use and production, as well as a new perception of society-nature interdependences. It is thus by combining a territorial approach with sustainability that a "sustainable territorial development model" can be configured to account for post-Fordist issues. The model is not mechanical and combines a partial approach accepting the incomplete and the unfinished (territory) with a globalizing, "ecosystemic" approach (sustainability). Its specific characteristics depend on such opposition.

As already stated, such multifaceted evolution will have clear impacts on the spatial structuring of national territories (DIACT 2009). An in-depth foresight analysis should enable the mutations taking pace to be better apprehended.

A Two-Pronged Analysis: the Territorial and The Sustainability Approach

The Territorial Approach: In Praise of Incompleteness, and the Discovery of Sustainability

In the emerging post-Fordist model, sustainability is seen as a key issue in resource renewal. It is through the concept of resource that sustainability and territory can be linked.

However, as noted by Olivier Godard (2007), it is not necessarily easy to identify territorial and sustainable development: "regional development – even supposing that "regional" and "territorial" are synonyms – may quite clearly not be sustainable."

The discussion of the territorial-sustainable pairing will continue by using the commonly agreed definitions for the words "territory" and "territorial."

The main feature of territorial resources is their specificity. The latter is a characteristic of a resource or production linked to a place, its history and its culture. It is the belief that what is made here could not be made anywhere else while keeping the same characteristics. The archetype of specific production could be the Eiffel Tower. It is completely linked to (dependent on) the city of Paris, and could not exist as such in any other place but Paris. Almost as clear an example is the quality labeling attached to certain European food products (AOP, IGP). In these cases, specificity is linked to a cultural product and a particular area of land (terroir), as well as to the area's history. It is a concept which gives a good idea of the value created by attaching a product to a territory. However, it overlooks another facet of specificity – its dependence on its territory. In other words, specificity is a territorial characteristic, which can be seen as positive in that it makes local production distinctive and allows it to free itself from normal standards, justifying the consumer's willingness to pay for it. It can also be seen in a negative light, however, as a form of dependence tying the activity to the place, thus preventing it from moving in order to reduce production costs.

The second characteristic of territorial-resource production is incompleteness. This is because the process of forming a territory consists in bringing the population of a given geographical context together (i.e., the context of a pre-existing lifestyle, culture and history) with the aim of voicing, then working to solve, a common problem. This constitutes a "small world" as defined by Zimmermann (2002), but not necessarily a complete production system. Changes in the way large corporations are organized in the production sphere are such that we need to go beyond the concept of "industrial districts" (following Marshall's use, as reinterpreted by the Italian economists Becattini et al. in 2009) or more mundanely "cluster." Thus, compared to traditional "local production systems" like those in rural France and Brazil (the present authors' homelands), territorial mergers are today more open. The Brazilian term for such systems, "Local Production Patterns," thus seem much more pertinent and suited to what is actually happening. Zimmermann (op. cit. p. 519) in particular notes that three important changes have occurred in the structure of production systems based on proximity (districts, clusters, LPSs, LPPs, local food production systems [SYAL], etc.): "It is not only small and middle-sized companies which set up local industrial systems but also ... local branches of large corporations. The coherence of such systems is not necessarily found in just their internal elements and resources, but also in their ability to make use of external resources. ... The fundamental dynamics of such systems lies as much in their ability to innovate as in their productive efficiency."

Today, mobility and globalization have made Industrial Districts something of an exception and a model difficult to generalize. Current tendencies are not pointing towards an increase in the numbers of such systems. In contrast, territorial dynamics have never been stronger. Incomplete forms of the organization of actors seem to lie ahead. Territories are shifting, open and provisional entities. They take on the external features of a time and place and can disappear. Such volatility and incompleteness are components of current modalities of the relationships between production, inhabitants and living areas. This involves supra-territorial interdependence and hence the question of the relationship between the incomplete part and the globalizing whole.

It thus follows that devising and implementing territorialized sustainable development will require the adoption of active solidarity with other territories as a basic principle (Vieira et al. 2006).

Sustainability as a "Regulatory Ideal": From the Disillusion of "Top-Down Macro" to the Charms of "Bottom-Up Development"

It is now clear that the biosphere is under considerable pressure due to increasing human impacts on ecosystems and landscapes: the interlocking complexity of worldwide pollution, accelerated loss of biodiversity (and sociodiversity), desertification, climate disruption, chaotic urbanization, social exclusion and endemic poverty. As these are neither mono- nor multidisciplinary macro-problems, they need to be looked at from an alternative standpoint considering all aspects of what is a co-evolutive "society-nature" process. Understanding the phenomenon does not mean juxtaposing isolated disciplinary factors, but rather exposing their interdependence at the core of a global system made up of the interactions between them. Rather than reduce things to this or that aspect by dividing up scientific disciplines into distinct compartments, systems-oriented research endeavors to take them as a whole and place them in their contexts. (Von Bertalanffy 1968; Passet 1979; Berkes et al. 2003; Morin and Kern 1993).

Opening up to a systemic-complex world-view and to the creation of new lifestyles leads to the rediscovery of potential, productive capacity, distinctive identities and the energy of endogenous local development initiatives in the face of the advancing effects of economic and cultural homogenization and deterritorialization brought on by Fordist rules of economic growth. Countering the model of excessive growth, with poverty affecting more than two thirds of the world's population in addition to intensive degradation of the Earth's natural heritage and human habitats, is the key to unfastening the Gordian knot in question.

In the follow up of the Stockholm Conference, held in 1972, interdisciplinary research efforts related to the endogenous dimension of development strategies were carried on by members of the Centre International de Recherche sur l'Environnement et le Développement in Paris, under the leadership of Ignacy Sachs (CIRED 1986). Using the method of a radical analysis and assessment of economism, the researchers avoided falling into the twin traps of technocratic statism or the inadequacy of a fragmented autarkical or localist approach to strategies of socio-economic and cultural regeneration. The environment was considered as a basic developmental dimension to be integrated in a new, unified and participatory planning and management approach.

The concept of local control over development choices was emphasized in order to point up communities' needs to seek solutions to their problems by relying on the knowledge, values and experience acquired in their own surroundings – which does not mean they should be left to fend for themselves with no outside contact. In this approach, asserting or preserving a local cultural identity means refusing development strategies resulting in greater dependence on outside authorities or organizations, while being compatible with belonging selectively and critically to wider political and cultural entities.

This concept of eco-development was initially presented as a heuristic, normative approach to planning – a sort of philosophy of development – and not as a new, fully developed theory (Sachs 1980, 1981; Dag Hammarskjöld Foundation 1975). Nevertheless, the concept was progressively extended and enhanced to put together a new theoretical and methodological corpus in the field of a systems-oriented human ecology (Vieira et al. 2005).

Twenty years later, under the influence of the Earth Summit, the debates on the concepts of Agenda 21 and sustainability gave rise to at least two very different interpretations of socio-ecological issues (Lévesque 2009). The first, called strong sustainability, holds that the guiding principles to shape new development strategies are descendants of the "classical" approach to eco-development. It refers to a sort of experimental harmonization game in which socio-economic objectives are co-related to the fulfillment of basic human needs, ecological prudence, political descentralization and self-reliance. Economy is reduced to its function as a simple set of means, "designed to achieve the goals of social justice and ecological sustainability whose content is the result of participative political decision-making" (Laville 2005, p. 336). According to this view, the environment is thus taken as an essential dimension to be internalized. The practice of sustainable development is related to strengthening non-commercial, non-monetary sectors connected to regulated markets" (ibid., p. 337). The strong interpretation tends to widen the scope of the extra-economic, the plural economy and even the mixed economy.

On the other hand, weak sustainability, is based on a formal rather than a substantive definition of the economy: the degradation of natural heritage can easily be counterbalanced by an increase in the volume of production and consumption, resulting in the paradox of a search for "an optimal rhythm of the destruction of nature!" (Marěchal 2005, p. 44). On the other hand, in the words of Brohman (2000, p. 313), issues of sustainable development in today's interconnected world cannot be coherently addressed ouside of their North-South context, especially the contradictions imposed by the structural inequalities of global capitalism.

During the 1980s, when national economies were adapting to neo-liberalism, the territorial approach became increasingly used to designate the local, e.g. in regard to the effects of proximity, to the endogenous and to territorial systems of governance. The desire to revitalize the endogenous as the main guiding principle of new development strategies shows "that there are ways of organizing social life and production rooted in territory, that is to say where the socio-cultural and historical contexts (the specific nature of a territory) predominate" (Gumuchian and Pecqueur 2007, p. 5). Territorializing the concept of local development makes clear "the efficiency of social relationships which are not exclusively commercial in making the most of the sources of wealth at people's disposal" (Pecqueur 1989, p. 17). From this point of view, territorial development turns out to be a new, combative means of adapting to globalization. It is a way for companies to get round the competition of price and production costs by concentrating on quality and on cooperative relationships between stakeholders (Pecqueur 2006). Innovation plays an essential part, allowing the capacity for rapid reaction to be emphasized and changes to be anticipated. The territorial approach prefers more community-based ways of working to cumbersome, hierarchical, pyramid-shaped processes, whose adaption to current conditions of global opportunity is becoming increasingly inadequate.

At the same time, Claude Courlet cautions that this approach tends to focus on a "meso-economy of territories", with their specific means: local production systems, innovatory settings, clusters, urban production systems, as well as concepts of territorial competitiveness, territorial resources, local governance etc. (Courlet 2008, p. 119). While acknowledging that "although many research studies on territory are being carried out, they seldom look at the question of complexity," he points out that" a territory has all the characteristics of a coherent, complex system" (ibid., p. 34).

In our view, several central aspects of the debate on territorial development strategies fall within the traditional sustainability framework put forward by the advocates of eco-development. Fresh inputs have opened up new theoretical possibilities, however. One example is the new local-global configurations brought about by globalized features taking root at the local level, another is analysis based on the concepts of proximity, territorial resource, territorial competitiveness, the knowledge-based economy, innovatory settings and territorial governance.

Such new analytical frameworks are now being hybridized, and cover areas as apparently diverse as current consumption and lifestyle patterns; the dynamics of economically-viable appropriation, use and management of natural resources and territories; technological risks; the promotion of equity and cooperation; and governance. These frameworks also question the supposed autonomy of the economic-activity cycle with respect to the state of the environment and other aspects of social life; they identify, among the factors currently involved in the worsening environmental crisis, different aspects of the crisis of the underlying organizational patterns of contemporary society (Vieira et al. 2010).

It is in this context that decentralization and "centralized synthesis" can be seen as complementary techniques to be mobilized in the field of political decision-making. Decentralization reinforces self-reliance – self-confidence, endogenous action, initiatives by local actors in the planning and management realms. This implies subordinating market economy to the constraints imposed by seeking the resilience of ecosystems and by understanding the factors enabling territorial identity and local and regional cultural life to survive (Berkes et al. 2003; Vieira et al. 2005). However, preserving cultural identity would not be sufficient in the long term. In our view, it is the ability to enhance economic activity and live in harmony with it, which seems crucial. On the other hand, "centralized synthesis" provides crucial assets for facing up to scientific uncertainty, to the more-or-less unforeseeable effects of the dynamics of socio-ecological systems, to the multiplicity of world views, socio-political ideologies and technological options (Godard and Sachs 1975; Sachs 2006).

Reflecting a new principle of social rationality, this new analytical framework could also allow for more coherent connections between three scales of intervention. Firstly, in the context of so-called cognitive ecology, involving a fairly radical rupture with our rigid way of perceiving the meaning of humanity's presence in the universe. Secondly, at the scale of global ecology, which provides us with an increasingly clear perception of the seriousness of global environmental change and of the importance of North-South asymmetry in shaping and reinforcing these trends. Lastly – in an attempt to connect the first two – in the intermediate scale of actions aiming to create integrated, decentralized, collaborative systems of managing socio-ecological systems.

Coherence, Convergence and Hybridization?

Territorial Resources Are Infinite… but Depend on the Environmental Context

Territorial resources are specific, meaning that they are created. They combine primary resources, some of which are fragile and non-renewable, i.e. likely to disappear either by running out or through being damaged, and immaterial resources such as culture and history. What constitutes the overall territorial resource is the transformation of these ingredients. A territorial resource is thus not simply a natural resource, but has undergone a cultural transformation to become a specific asset.

There is thus a close association between immaterial resources and the issue of sustainability. When associated with a place, specific resources become infinite and linked to the notions of quality and renewability.

The infinite nature of this type of resource can be shown by the observation that specificity can depend on some slight detail. The case of Dinant, a town in Belgium, is a good illustration: the local music industry (instrument-making, the organization of music schools and concerts, etc.) has created not only jobs but also a strong economic momentum. At the origin of this specificity is a simple biographical incident – the fact that Adolphe Sax, the inventor of the saxophone, was born in the town.

Potential reproducibility is dependent on the type of specificity. The concept of specificity is inseparable from that of quality. Whether it is an intrinsic quality linked to the product itself and/or the way it is made, or an extrinsic or subjective quality linked to the reputation and image of a place (especially to the landscape factor), specificity depends on the conditions for renewing resources. It can thus be seen that even if territorial resources are not directly drawn from renewable resources, they are directly dependent on them.

Sustainability: Managing the Global Resource Through a Systemic Approach

In the approach to sustainable territorial development outlined here, the questions of contingent uncertainty, constraints of viability and scientific controversies are provocative points for highlighting the limits of prospective scenarios in systems of socio-environmental management (Vieira and Weber 2000). Such points reflect the tensions and paradoxes associated with the evolving dynamics of complex macrosystems and are seen in the fields of biological, cognitive and even social sciences. A priori, management systems need to be ready to withstand a variety of unpredictable disturbances and fluctuations while being ready to reorganize themselves in such situations via adaptive learning strategies still in their first stages of development (Armitage 2007; Berkes 2009; Olsson et al. 2004). The resilience of socio-environmental systems is seen here as the essential condition for implementing the concept of strong sustainability (Holling 1978, 1998, 2001).

This brings us not only to the question of the economic status as such of collective assets, but also to the conflicts of representation resulting from the involvement of a wide range of social actors, in addition to the scientific uncertainties and controversies concerning the conditions in which ecosystems and landscapes are reproduced over long time spans.

Meanwhile, the search for a common legal status for the integrated, collaborative management of socio-ecological conflicts has unearthed an important innovative criterion in the concept of natural and cultural heritage, against the historical backdrop of a development approach considered inherently contradictory in socio-ecological terms (Vivien 1994). As François Ost has insightfully noted (1995, p. 351), this concept helps endow the ethical concern for the chances of survival of present and future generations with an increasingly convincing legal basis. To quote:

> Relationships between humans and the environment do not fit in well with the statuses of object and subject. It is as if the distinction between subject and object, on which our modernity has been so solidly built, were completely at odds with the need to think through and manage an interactive reality like the environment. At the same time, the limitations constitutive of legal approaches to the same reality have become visible, whether as limitations possible to express in terms of appropriation, contract or regulation, or, on the contrary, as propositions linked to the personification of nature and the recognition of its rights. Finally, it is the distinction between the public and private domains, between public and private law, which needs to be overcome if suitable responses are to be made to environmental issues.

Finally, the innovatory, synergetic responses emerging in some areas of local self-organization have begun to catch the attention of researchers in search of a new planet-wide civil society: they are places "where new social movements are spreading, where unprecedented functions and structures are being established, where new relationships between people and nations are being invented, where the world and society are being theorized outside the fixed canons of dominant orthodoxy and its usual refutations" (Ziegler 2002, p. 283). Despite the constraints imposed by globalized markets and trade, current empirical evidence confirms that various cooperative action groups are succeeding in setting up local and regional mutual-aid networks in an effort to break out of their isolation and to avoid competing with one another on the open market. In places where socio-economic vigor seems to stem from new structural, hybridized modes of several types of economic activity (commercial, non-commercial and non-monetary), it is turning out to be possible to develop the latent or underused potential of natural and cultural resources.

The issue of "global change" thus runs counter to the process of passive adjustment to the constraints engendered by the current dynamics of globalization. It is an issue, which in the last few years has become an international beacon transcending all localities and cultures, whether in the North or the South. In our view, there are two major concerns at the heart of this phenomenon: one is ever-spreading inequality due to mechanisms for depriving people of resources and rights, and the other is the worsening ecological crisis due to global warming and the growing greenhouse effect. New cooperative practices are emerging based on renewed perceptions of development issues (Tremblay and Vieira 2012).

It is no longer a question of focusing solely on socio-economic aspects by making employment the cornerstone of discourse on living together at local scales. If the necessity for social inclusion by creating productive jobs is indisputable – especially in rural areas – an effort must also be made to "get back to basics" and make more obvious the importance of the organic links between environmental aspects and socio-economic, socio-cultural and socio-political ones in the search for effective ways out of the global crisis.

The emergence of new, supportive, ecologically responsible practices implies that further thinking is required on the conditions in which sustainable territorial development strategies can be viable. In our opinion, such practices reveal the direction in which more substantial theoretical, methodological and ethical field research and action should go in view of the challenges thrown up by the "era of catastrophes towards which we are now irreversibly hurtling" (Stengers 2009; Dupuy 2002). This will involve both the approach to and theoretical treatment of local development strategies and initiatives set up or actively supported by public, business or not-for-profit stakeholders based on the concept of territorial development.

The complexity of socio-ecological interactions clearly makes it difficult to include environmental issues in conventional development-planning and -management structures. The lack of widely-accepted scientific knowledge of ecosystem and human-influenced landscape processes is an important consideration, as it would be completely contradictory to treat environmental issues in a piecemeal, compartmentalized way, while leaving intact the reductionist conceptual framework and established procedures. The approach outlined here requires defining "territory" as a complex, open system whose socio-economic, socio-cultural and socio-political interactions are to be studied and the specific features of its biophysical and built environments taken into account. To modify the system, the priority lies in understanding the horizontal (inter-sector) and vertical (between regulatory levels) networks of interrelationships evolving over time in a non-linear manner via positive and negative retroaction loops and "system effects" (Ostrom 2002; Gunderson and Holling 2002; Cash et al. 2006).

The "Quality Model" (or Sustainable Territorial Development)

The term "inter-territorial competition" may suggest that such competition takes place between territories either presumed to exist or actually existing, i.e. combining a specific resource with an identified, spatially-defined territory.

As we hope we have made clear, the "spatial competition factors" we propose to adopt are based on a different perception of the process of territorial construction. A new territorial model thus means taking a fresh look at the idea of the division of labor based on Ricardo's "comparative advantage" and focusing on the decisive questions of coordinating and creating resources.

A model of territorial quality or sustainable territorial development should therefore combine the creation of a "differential" advantage or specificity, of intrinsic or extrinsic quality and of the ability to renew resources.

We propose below a selection of stakeholder behaviors according to two models, which although overstated sum up the issues at stake in the present discussion.

The Table 1 highlights the fact that the strategy of adapting to globalization by placing oneself at the core of competition (with the traditional advantages of competing through cost) does not pay off. Indeed, the major characteristic of territorial

Table 1 Stakeholder behaviors

Productivity model:	Territorial quality model:
1. **Endowment** of factors to optimize	**Resources** to build up
2. Low costs and low prices:	Stable high prices:
Constrained by **production costs**	Constrained by **quality labels**
3. **Individualized** productivity	**Globalized** productivity
4. Generic products **(differentiation)**	Specific products **(specification)**
5. Global governance (following **corporate logic**)	Local governance (following multi-dimensional **stakeholder logics**)
6. **Competition** as the dominate force	Strategy of **avoiding competition**
7. **Corporate** logic	**Territorial** logic
8. **Profit**	**Secure income**
9. **Exogenous** innovation	**Endogenous** innovation
10. **Distinction** between public and private ownership	**Combination** of public and private ownership
11. The environment is a **constraint**	Harmonizing the ecological, social and economic dimensions of development
12. **Setting a price** for exploitation **independent of** the constitution of the resource	**Managing** exploitation **as part of** the constitution of the resource

development lies in the attempt to extricate oneself from competition by establishing a secure income. This type of strategy relies on product specificity, i.e. products defined by their territory. This can be achieved in several ways: the most obvious one is to use origin as a discriminating factor; cultural heritage or landscape can also be used as territorial production indicators, as can knowhow built up and accumulated via the processes of constituting a "collective cognitive memory," etc.

We are therefore left, in the empirical real world, with a combination of two modes of development, where labor productivity – Fordism's main driving force – is counterbalanced by the ability of territories to distinguish themselves through specificity and quality – including the need to integrate the pursuit of social inclusion and ecological prudence.

Conclusions

The sustainable territorial development approach combines the questions of the relative autonomy of local dynamics and the interconnectedness of different territorial scales with radical changes in the relationships taking place between economic actors, the State and civil-society organizations. Most of the problems societies are currently facing are too complex to fit into traditional "top-down" government structures any longer. Whereas public policies are devised within administrative and sectorial frameworks, inherently trans-scalar socio-environmental problems clearly go beyond such boundaries. That is why the viability of this new development concept in a time of worsening global crisis is indicative of the usefulness (among

(continued)

other socio-economic and socio-cultural conditions) of the key concept of territorial governance. This consists of a process producing public policies by multi-actor structures not included in formal politico-administrative hierarchies (Carlsson and Sandström 2008).

Two decades after the first Earth Summit, however, where should realistic options be sought for firmly implanting such socio-technical innovations? To what extent are contemporary transformations in the running of whole societies and regional and local communities really favorable towards the sustained, concerted experimentation of alternatives based on the promotion of territorial dynamics and sustainable development? In what way is the latter concept really useful in dealing with the world socio-ecological crisis? How, moreover, can we determine and analyze the complex inter-dependence of the issues at stake and bring about a dialogue of knowledge linked to the need for the clearheaded sharing of experience – in the sense of the empowerment of local communities – so as to create a source of real commitment manifesting itself in real changes in the perceptions, attitudes and behavior of each of its constituent actors? These are extremely difficult questions, which should rapidly encourage us to make up for the many inadequacies in applying sociologically oriented acquirements of human ecology to the coordinated construction of sustainable territories in the years ahead.

By judiciously parsing the traditional approach to eco-development, the concept of sustainable territorial development put forward in this chapter could become a hybrid construction, grounded on the expertise of systemic-complex thinking applied to the environment-development nexus and enhanced by the more recent insights of the groundbreaking school of territorial economics.

Such ideas and propositions remain controversial in the contemporary geopolitical situation. They need to be refined and validated by rigorous comparative case studies, in an intellectual setting favorable to experimental creativity. It is nevertheless our hope that the sustainable territorial development approach can lay down credible paths for advancing towards initiatives seeking to get beyond the traditional view of homo oeconomicus stuck in the dominant paradigm of human development – with all that such an approach may imply of uncertainty and of transitory hypothetical propositions.

The strategic task to be fulfilled is nothing less than designing new forms of community living, in which whatever is possible will be considered in the light of principles for the global management of the biosphere. As technology is now the most powerful agent of social change, the deciding battles will be won or lost according to how seriously we take the major challenge lying before us, to wit: promoting the metamorphosis of mixed systems of socio-technological realities so that they correspond to new strategies for creating and developing resources which territories can benefit from.

References

Armitage D, Berkes F, Doubleday N (2007) Adaptive co-management: collaboration, learning, and multi-level governance. University of British Columbia Press, Vancouver
Becattini G, Bellandi M, De Propis L (2009) A handbook of industrial districts. Edwrd Elgar, Cheltenham
Berkes F (2009) Evolution of co-management: role of knowledge generation, bridging organizations and social learning. J Environ Manag 90:1692–1702
Berkes F, Colding J, Folke C (2003) Navigating social-ecological systems: building resilience for complexity and change. Cambridge University Press, Cambridge
Bouba-Olga O (2006) Les nouvelles géographies du capitalisme, comprendre et maîtriser les délocalisations. Éditions Le Seuil, Paris
Bourg D, Whiteside K (2010) Vers une démocratie écologique, le citoyen, le savant et le politique. La république des idées, Le Seuil, Paris
Boyer R (2004) Théorie de la régulation, 1. Les fondamentaux, Editions La Découverte, Paris
Brohman J (2000) Popular development. Rethinking the theory and practice of development. Blackwell Publishers, Oxford
Carlsson L, Sandström A (2008) Network governance of the commons. Int J Commons 2:33–54
Cash DW, Adger NW, Berkes F, Garden P, Lebel L, Olsson P, Pritchard L, Young O (2006) Scale and cross-scale dynamics: governance and information in a multilevel world. Ecol Soc 11(2):Article 8
CIRED (1986) Rapport scientifique 1973–1986. EHESS, Paris
Colletis G, Paulré B (2008) Les nouveaux horizons du capitalisme, pouvoirs, valeurs, temps. Editions Economica, Paris
Courlet C (2008) L'économie territoriale. Presses Universitaires de Grenoble, Grenoble
Dag Hammarskjöld Foundation (1975) Que faire. DHF, Uppsala
Dupuy J-P (2002) Pour un catastrophisme éclairé. Le Seuil, Paris
Gille B (1978) Histoire des techniques, Encyclopédie de La Pléiade. Gallimard, Paris
Godard O (2007) Du développement régional au développement durable: tensions et articulations. In: Mollard A, Sauboua E, Hirczak M (eds) Territoires et enjeux du développement régional. Quae, Paris, pp 83–96, Update sciences et technologies
Godard O, Sachs I (1975) L'environnement et la planification. In: Barreau M (ed) Environnement et qualité de la vie. Guy le Prat, Paris, pp 207–247
Gumuchian H, Pecqueur (2007) La ressource territoriale. Economica, Paris
Gunderson L, Holling CS (2002) Panarchy: understanding transformations in human and natural systems. Island Press, Washington, DC
Holling CS (1978) Adaptive environmental assessment and management. Wiley, New York
Holling CS (2001) Understanding the complexity of economic, ecological and social systems. Ecosystems 4:390–405
Holling CS, Folke C (1998) Science, sustainability and resource management. In: Berkes F, Folke C (eds) Linking social and ecological systems. Cambridge University Press, Cambridge, pp 342–362
Klein JL (2008) Territoire et développement, du local à la solidarité interterritoriale". In: Massicotte G (ed) Sciences du Territoire. Presses Universitaires du Québec, Massicotte, pp 315–333
Laville JL (2005) Solidarité et développement durable. In: Maréchal JP, Quenault B (eds) Le développement durable. Presses Universitaires de Rennes, Rennes
Le Cercle des économistes (2009) Mobiliser les territoires pour une croissance harmonieuse. Rapport officiel pour le Président de la République, La Documentation Française, Paris
Lévesque B (2009) Economia plural e desenvolvimento territorial na perspectiva do desenvolvimento sustentável. Política Soc 8(14):107–144
Maréchal JP (2005) De la religion de la croissance à l'exigence de développement durable. In: Maréchal JP, Quenault B (eds) Le développement durable. Presses Universitaires de Rennes, Rennes

Moati P (2001) L'avenir de la grande distribution. Editions Odile Jacob, Paris
Morin E, Kern AB (1993) Terre Patrie. Le Seuil, Paris
Moulier-Boutang Y (2007) Le capitalisme cognitif, la nouvelle grande transformation. Editions Amsterdam, Paris
Olsson P, Folke C, Berkes F (2004) Adaptive comanagement for building resilience in social-ecological systems. Environ Manag 34(1):75–90
Ost F (1995) La nature hors la loi. La Découverte, Paris
Ostrom E, Dietz T, Dolsak N, Stern PC, Stonich S, Weber EU (2002) The drama of the commons. National Academic Press, Washington, DC
Passet R (1979) L'économique et le vivant. Payot, Paris
Pecqueur B (1989) Le développement local. Syros, Paris
Pecqueur B, Zuindeau B, (2010) "Espace, territoire, développement durable," Développement durable et territoires, Zuindeau B. ed., pp 49–56
Sachs I (1980) Stratégies de l'écodéveloppement. Les Éditions Ouvrières, Paris
Sachs I (2006) Rumo à ecossocioeconomia. Teoria e prática do desenvolvimento. Cortez Editora, São Paulo
Sachs I et al (1981) Initiation à l'écodéveloppement. Privat, Toulouse
Sassen S (2006) Territory, authority, rights: from medieval to global assemblages. Princeton University Press, Princeton
Stengers I (2009) Au temps des catastrophes, Résister à la barbarie qui vient. La Découverte, Paris
Talandier M, Davezies L, (2009) Repenser le développement territorial? (Confrontation des modèles d'analyse et des tendances observées dans les pays développés), Editions du PUCA, série Recherche, Paris
Tremblay G, Vieira PF (2012) Le rôle de l'université dans le développement local. Expériences brésiliennes et québécoises. Presses Universitaires du Québec, Québec
Vanier M (2008) Le pouvoir des territoires, Essai sur l'interterritorialité. Editions Economica, Paris
Veltz P (2009) Le nouveau monde de production. Sci Hum 210:46
Vieira PF, Weber J (2000) Gestão de recursos naturais renováveis e desenvolvimento: novos desafios para a pesquisa ambiental, 2ª edn. Cortez Editora, São Paulo
Vieira PF, Seixas CS, Berkes F (2005) Gestão integrada e participativa de recursos naturais: conceitos, métodos e experiências. APED, Florianópolis
Vieira PF, Cazella A, Cerdan C, (2006) Desenvolvimento territorial sustentável. Conceitos, experiências e desafios teórico-metodológicos. Eisforia, vol. 4, n°4
Vieira PF, Cazella A, Cerdan C, Carrière J-P (2010) Desenvolvimento territorial sustentável no Brasil. Subsídios para uma política de fomento. APED/SECCO, Florianópolis
Vivien FD (1994) Économie et écologie. La Découverte, Paris
Von Bertalanffy L (1968) General systems theory. Braziller, New York
Zaoual H (2005) Socioéconomie de la proximité: du global au local. L'Harmattan, Paris
Ziegler J (2002) Les nouveaux maîtres du monde et ceux qui leurs résistent. Fayard, Paris
Zimmermann J-B (2002) "Grappes d'entreprises" et "petits mondes," une affaire de proximités. La Revue Economique 53(3):517–524

Rheims Sustainability Vision

A contribution to the Open Working Group on Sustainable Development Goals, made as the conclusion of the Third *Rencontres Internationales de Reims* on Sustainability Studies on the Sustainable Development Goals and the post-2015 Development Agenda.

Building on the Rio+20 United Nations Conference on Sustainable Development, its outcome document *The Future We Want* and the post-2015 Development Agenda, particularly regarding Sustainable Development Goals,

Seeking to contribute to the Open Working Group on Sustainable Development Goals,

Believing that the new social contract represented by the future Sustainable Development Goals should apply universally, regardless of a person's citizenship or place of residence,

Emphasizing the importance that some basic ethical principles underlie this new social contract,

Recognizing the need to identify resistances to social and economic justice and the concrete means to overcome them, as declarations of ideals are not sufficient,

Recognizing the importance of facilitating local and regional adaptations of global sustainable development principles and goals,

Emphasizing the role of regional organizations, alongside global and national actors, in promoting and supporting the implementation of sustainable development goals,

Reminding the international community about the impact of globalization and the role of multinational corporations, as it will be impossible to get rid of poverty, address malnutrition, deal with greenhouse gas concentrations without significant changes to the international corporate space,

Reminding in particular about the role of large media corporations and about the importance of their accountability to make key sustainability issues visible to and understood by the larger public,

F. Mancebo, I. Sachs (eds.), *Transitions to Sustainability*,
DOI 10.1007/978-94-017-9532-6

Reaffirming that the Sustainable Development Goals are not only about setting goals, but also about identifying a legitimate process to reach them and that the following aspects of the process need careful attention:

- participation and dialogue (how can communities be engaged in a way that empowers them to identify their own goals and development pathways among and beyond the goals),
- transparency and accountability (how governance process on issues such as energy that are traditionally confined to small groups of closed networks be opened up and made transparent and accountable towards those whose lives their decisions influence),
- equity and fairness (how can gender equity be achieved in the governance processes around Sustainable Development Goals, particularly in areas where they are in considerable minority such as energy),

Considering that, while there must be room for a sectorial approach, integrated and overarching goals would go a long way in addressing the environmental challenges the world is facing; for example, providing universal access to services that are essential for the survival and development of an individual, including water, energy, food and nutrition and health care; protecting the rights of all vulnerable groups, especially women, children and minorities, through the development, for example, of an aspirational model rights framework, which should recognize issues of culture and national diversity and allow for exceptions to specific provisions, and moving towards maximizing resource efficiency,

Being aware of the importance of resilience of mountain, arctic, island, coastal and other particularly vulnerable regions to global environmental change such as global warming and, hence, the importance of paying special attention to fostering and monitoring sustainable development in these areas,

Recalling the focus on planning of the First *Rencontres Internationales de Reims* on Sustainability Studies in 2011 and the focus on governance of the Second Rencontres in 2012,

The participants in the Third *Rencontres Internationales de Reims* on Sustainability Studies, held in Reims on 19–20 June 2013, recommend to

- Develop goals that have a sound scientific base and strong inner consistency in order to ensure that they are truly sustainable and not just a random list of priorities for the international community;
- Develop indicators of development that address the facilitating conditions for every person to rise out of poverty;
- Adopt targets and indicators for sustainable lifestyles that respect planetary resource limits and equity;
- Promote access to information and justice, supported by transparent subsidies if necessary, to ensure broad and meaningful public participation in decision-making;
- Reinforce democratic and participatory institutions as key for advancing sustainable development integrating the social, economic and environmental dimension;

- Reinforce development goals and augment them with targets and indicators of sustainability;
- Restructure the energy sector in a manner that energy is provided sustainably from sources that do not emit greenhouse gases;
- Ensure that energy has a place of its own among Sustainable Development Goals through an overarching goal of sustainable energy for all, as recommended by many: while energy access and consumption is not an end in itself, it is a means to many ends, as many Millennium Development Goals in the field of poverty, health and environment are intimately linked to the way energy is produced and consumed;
- Ensure that a separate Sustainable Development Goal is dedicated to water, which is essential for life and has already been recognized as a human right, and that the misunderstandings, which led water to be dropped from Millennium Development Goals and from the outcome document of Rio+20, are overcome using existing international legal instruments to overcome these misunderstandings;
- Ensure that education also has a strong presence in the framework, not only with quantitative goals, but also with qualitative ones, ensuring that education is built on values that unify rather than divide, that integrate rather than separate and that build capacities of individuals to serve their local and global community;
- Ensure that a specific target on access to health services and a healthy environment is set because of its importance for sustainability;
- Create mechanisms so that all corporate and private wealth and wealth creation around the world contribute their fair share to tax revenues to support collective services and the public good;
- Require that all multinational enterprises publish a sustainable development report integrated with their financial reports or explain why they are not doing so;
- Valorize and compensate material and non-material ecosystem goods and services, particularly those originating in mountain, arctic, island, coastal and other environmentally valuable areas;
- Enhance disaster risk reduction and preparedness in climate change-threatened upland, lowland and coastal areas and secure biodiversity corridors along altitudinal gradients;
- Improve communication infrastructure, including access to broadband Internet, in remote regions, especially mountain, islands and other remote and less inhabited areas to overcome the digital divide;
- Envisage goals for implementation at the community level through solidarity and empowerment;
- Reinforce our thinking and dialogue on the content and substance of possible alternative development goals for the long term, at the community, national and international levels as well as on new methods of planning aimed at translating broad objectives into practical, realistic and effective strategies;
- Create a Global Sustainability Panel to provide expertise for supporting the achievement of the Sustainable Development Goals, to assess and identify

urgent sustainability challenges and to coordinate the multiple international science panels already providing valuable knowledge and epistemic support for ongoing efforts at environmental protection at the global level.

The following individuals contributed to the preparation of this document:

Liliana Andonova (Geneva, Switzerland), Frank Biermann (Amsterdam, Netherlands), Jörg Balsiger (Geneva, Switzerland), Jon Church (Rheims, France), Christian Comeliau (Geneva, Switzerland), Arthur Dahl (Geneva, Switzerland), Marc Dijk (Maastricht, Netherlands), Ladislau Dowbor (São Paulo, Brazil), Harris Gleckman (Boston, USA), Peter Haas (Amherst, USA), Sylvia Karlsson-Vinkhuyzen (Wagenigen, Netherlands), Carlos Lopes (Addis Ababa, Ethiopia), François Mancebo (Rheims, France), Alexander Mejia (Geneva, Switzerland), Thomas Perianu (Paris, France), Carlo Rubbia (Potsdam, Germany), Ignacy Sachs (Paris, France), Leena Srivastava (New Delhi, India).

Rheims, June 20, 2013

MIX
Papier aus verantwortungsvollen Quellen
Paper from responsible sources
FSC® C105338

If you have any concerns about our products,
you can contact us on
ProductSafety@springernature.com

In case Publisher is established outside the EU,
the EU authorized representative is:
Springer Nature Customer Service Center GmbH
Europaplatz 3, 69115 Heidelberg, Germany

Printed by Libri Plureos GmbH
in Hamburg, Germany